Hydrogen Safety

GREEN CHEMISTRY AND CHEMICAL ENGINEERING

Series Editor: Sunggyu Lee
Ohio University, Athens, Ohio, USA

Proton Exchange Membrane Fuel Cells: Contamination and Mitigation Strategies
Hui Li, Shanna Knights, Zheng Shi, John W. Van Zee, and Jiujun Zhang

Proton Exchange Membrane Fuel Cells: Materials Properties and Performance
David P. Wilkinson, Jiujun Zhang, Rob Hui, Jeffrey Fergus, and Xianguo Li

Solid Oxide Fuel Cells: Materials Properties and Performance
Jeffrey Fergus, Rob Hui, Xianguo Li, David P. Wilkinson, and Jiujun Zhang

Efficiency and Sustainability in the Energy and Chemical Industries: Scientific Principles and Case Studies, Second Edition
Krishnan Sankaranarayanan, Jakob de Swaan Arons, and Hedzer van der Kooi

Nuclear Hydrogen Production Handbook
Xing L. Yan and Ryutaro Hino

Magneto Luminous Chemical Vapor Deposition
Hirotsugu Yasuda

Carbon-Neutral Fuels and Energy Carriers
Nazim Z. Muradov and T. Nejat Veziroğlu

Oxide Semiconductors for Solar Energy Conversion: Titanium Dioxide
Janusz Nowotny

Lithium-Ion Batteries: Advanced Materials and Technologies
Xianxia Yuan, Hansan Liu, and Jiujun Zhang

Process Integration for Resource Conservation
Dominic C. Y. Foo

Chemicals from Biomass: Integrating Bioprocesses into Chemical Production Complexes for Sustainable Development
Debalina Sengupta and Ralph W. Pike

Hydrogen Safety
Fotis Rigas and Paul Amyotte

GREEN CHEMISTRY AND CHEMICAL ENGINEERING

Hydrogen Safety

Fotis Rigas

Paul Amyotte

CRC Press is an imprint of the
Taylor & Francis Group, an **informa** business

CRC Press
Taylor & Francis Group
6000 Broken Sound Parkway NW, Suite 300
Boca Raton, FL 33487-2742

Printed in the United States of America on acid-free paper
Version Date: 20120531

International Standard Book Number: 978-1-4398-6231-5 (Hardback)

Visit the Taylor & Francis Web site at
http://www.taylorandfrancis.com

and the CRC Press Web site at
http://www.crcpress.com

Contents

Series Preface

The subjects and disciplines of chemistry and chemical engineering have encountered a new landmark in the way of thinking about developing and designing chemical products and processes. This revolutionary philosophy, termed *green chemistry and chemical engineering*, focuses on the designs of products and processes that are conducive to reducing or eliminating the use and/or generation of hazardous substances. In dealing with hazardous or potentially hazardous substances, there may be some overlaps and interrelationships between environmental chemistry and green chemistry. Whereas environmental chemistry is the chemistry of the natural environment and the pollutant chemicals in nature, green chemistry proactively aims to reduce and prevent pollution at its very source. In essence, the philosophies of green chemistry and chemical engineering tend to focus more on industrial application and practice rather than academic principles and phenomenological science. However, as both a chemistry and chemical engineering philosophy, green chemistry and chemical engineering derive from and build on organic chemistry, inorganic chemistry, polymer chemistry, fuel chemistry, biochemistry, analytical chemistry, physical chemistry, environmental chemistry, thermodynamics, chemical reaction engineering, transport phenomena, chemical process design, separation technology, automatic process control, and more. In short, green chemistry and chemical engineering are the rigorous use of chemistry and chemical engineering for pollution prevention and environmental protection.

The Pollution Prevention Act of 1990 in the United States established a national policy to prevent or reduce pollution at its source whenever feasible. And adhering to the spirit of this policy, the Environmental Protection Agency (EPA) launched its Green Chemistry Program in order to promote innovative chemical technologies that reduce or eliminate the use or generation of hazardous substances in the design, manufacture, and use of chemical products. Global efforts in green chemistry and chemical engineering have recently gained a substantial amount of support from the international communities of science, engineering, academia, industry, and government in all phases and aspects.

Some of the successful examples and key technological developments include the use of supercritical carbon dioxide as a green solvent in separation technologies; application of supercritical water oxidation for destruction of harmful substances; process integration with carbon dioxide sequestration steps; solvent-free synthesis of chemicals and polymeric materials; exploitation of biologically degradable materials; use of aqueous hydrogen peroxide for efficient oxidation; development of hydrogen proton exchange membrane (PEM) fuel cells for a variety of power generation needs; advanced biofuel productions; devulcanization of spent tire rubber; avoidance of the use of chemicals and processes causing generation of volatile organic compounds (VOCs); replacement of traditional petrochemical processes by microorganism-based bioengineering processes; replacement of chlorofluorocarbons (CFCs) with nonhazardous alternatives; advances in design of energy-efficient processes; use of clean, alternative, and renewable energy sources in manufacturing; and much more. This list, even though it is only a partial compilation, is undoubtedly growing exponentially.

This book series on Green Chemistry and Chemical Engineering by CRC Press/Taylor & Francis is designed to meet the new challenges of the twenty-first century in the chemistry and chemical engineering disciplines by publishing books and monographs based on cutting-edge research and development to the effect of reducing adverse impacts on the environment by chemical enterprise. In achieving this, the series will detail the development of alternative sustainable technologies that will minimize the hazard and maximize the efficiency of any chemical choice. The series aims at delivering the readers in academia and industry with an authoritative information source in the field of green chemistry and chemical engineering. The publisher and its series editor are fully aware of the rapidly evolving nature of the subject and its long-lasting impact on the quality of human life in both the present and future. As such, the team is committed to making this series the most comprehensive and accurate literary source in the field of green chemistry and chemical engineering.

Sunggyu Lee

Preface

This book on the safe handling and use of hydrogen is intended as much to provoke thought and dialogue as to provide practical guidance. We of course hope that readers will find the information we present to be helpful in their endeavors to bring the envisaged hydrogen economy into reality. But we also hope that important areas of safety research, often viewed as being nontechnological, will be welcomed into the discussion of hydrogen safety along with the more familiar technological topics.

The book is organized to first address questions associated with the hazards of hydrogen and the ensuing risk in its use within industry and by the public. What are the properties of hydrogen that can render it a hazardous substance? How have these hazards historically resulted in undesired incidents? How might these hazards arise in the storage of hydrogen and with its use in vehicular transportation?

We then turn to the issues of inherently safer design and, in accordance with the previous comment on so-called nontechnological topics, safety management systems and safety culture. The European Commission (EC) Network of Excellence for Hydrogen Safety, *HySafe*, is singled out for separate coverage, as are various case studies associated with hydrogen and constructional materials. We conclude with a brief look at future research requirements and current legal requirements for hydrogen safety.

Our approach, then, has been to attempt a balanced view of hydrogen safety. Such a perspective comes not only from our respective areas of expertise, but also a collective belief that the safety of any material or activity is most effectively addressed through a combination of the physical sciences and engineering principles together with the management and social sciences. Non-hydrogen-related industries have undergone painful experiences in this regard; there should be no need for industrial applications involving hydrogen to experience the same difficulties.

Coauthor Fotis Rigas would like to acknowledge and thank his family, and especially his wife, Betty, for their (almost) uncomplaining and enduring patience during his long-drawn-out academic activities. He also wishes to express his gratitude to Ms. Allison Shatkin, senior editor in Materials Science and Chemical Engineering of CRC Press/Taylor & Francis, for entrusting him and coauthor Paul Amyotte with the task of writing this book.

Coauthor Paul Amyotte would like to acknowledge and thank his family for their loving support in all his professional undertakings. He is especially grateful to his wife, Peggy. He also extends gratitude to Dalhousie University, and in particular Dr. Joshua Leon, dean of Engineering, for his recent appointment as C. D. Howe Chair in Engineering.

The Authors

Fotis Rigas is an associate professor of Chemical Engineering at National Technical University of Athens in Greece and has been a Visiting Professor at National Autonomous University of Mexico. His current research and academic activities are in the areas of process safety and bioremediation of contaminated sites. He has published or presented over 150 papers in the fields of his activities and is a reviewer of papers in 34 international journals.

Paul Amyotte is a professor of Chemical Engineering and the C. D. Howe Chair in Engineering at Dalhousie University in Canada. His research and practice interests are in the areas of inherent safety, process safety, and dust explosion prevention and mitigation. He has published or presented over 200 papers in the field of industrial safety and is the editor of the *Journal of Loss Prevention in the Process Industries.*

List of Acronyms

ADNR (inland waterways)	Regulation for the Carriage of Dangerous Substances on the Rhine (EU)
ADR (road)	**A**ccord Européen Relatif au Transport International des Marchandises **D**angereuses par **R**oute - European Agreement Concerning the International Carriage of Dangerous Goods by Road
AIChE	American Institute of Chemical Engineers
ALARP	As Low As Reasonably Practicable Principle
ATEX	Appareils destinés à être utilisés en ATmosphères EXplosives (EU Directive)
BAMs	Bulk Amorphous Materials
BLEVE	Boiling Liquid Expanding Vapor Explosion
CCPA	Canadian Chemical Producers' Association
CCPS	Center for Chemical Process Safety
CEI	Dow Chemical Exposure Index
CFCs	Chlorofluorocarbons
CFD	Computational Fluid Dynamics
CL	Checklist
CNG	Compressed Natural Gas
COD	Code of Practice
CPU	Central Processing Unit
CSB	Chemical Safety Board (USA)
CSChE	Canadian Society for Chemical Engineering
CVCE	Confined Vapor Cloud Explosion
DDT	Deflagration-to-Detonation Transition
DGR	IATA Dangerous Goods Regulations
DNR	Department of Naval Research
DOE	United States Department of Energy
EIGA	European Industrial Gases Association
EPR	European Pressure Reactor
ETA	Event Tree Analysis
FDS	Fire Dynamics Simulator Code
F&EI	Dow Fire and Explosion Index
FMCSA	Federal Motor Carrier Safety Administration
FMEA	Failure Modes and Effects Analysis
FTA	Fault Tree Analysis

FVM	Finite Volume Method
GH_2	Compressed Hydrogen Gas
HAZOP	Hazard and Operability Analysis
HIAD	European Hydrogen Incident and Accident Database
HRAM	Hydrogen Risk Assessment Method
HSE	Health and Safety Executive
HTHA	High-Temperature Hydrogen Attack
IATA	International Air Transport Association
IChemE	Institution of Chemical Engineers (UK)
IEA	International Energy Agency
IGC	Industrial Gases Council
IMO (sea)	International Maritime Organization
ISD	Inherently Safer Design
ISO	International Standards Organization
ITER	International Thermonuclear Experimental Reactor
LBLOCA	Large Break Loss of Coolant Accident
LDL	Lower Detonability Limit
LFL	Low Flammability Limit
LH_2	Liquid Hydrogen
LLNL	Lawrence Livermore National Laboratory
LNG	Liquefied Natural Gas
LPG	Liquefied Petroleum Gas
MH	Metal Hydride
MHIDAS	Major Hazard Incident Database Service
MIACC	Major Industrial Accidents Council of Canada)
MIC	Methyl Isocyanate
MIL-STD 882	Military Standard 882
MOC	Management of Change
MOFs	Metal-Organic Frameworks
MVFRI	Motor Vehicle Fire Research Institute
NASA	National Aeronautics and Space Administration
NBP	Normal Boiling Point
NFPA	National Fire Protection Association
NHTSA	National Highway Traffic Safety Administration
NIST	National Institute of Standards and Technology
NRA	National Railway Authority
NTP	Normal Temperature and Pressure
NWC	Naval Weapons Center
OHA	Operating Hazard Analysis
OHSAS	Occupational Health and Safety Assessment Series

OSHA	U.S. Occupational Safety and Health Administration
PAR	Passive Auto-Catalytic Recombiners
PED	European Pressure Equipment Directive
PHA	Preliminary Hazard Analysis
PRD	Pressure Relief Devices
PSM	Process Safety Management
QRA	Quantitative Risk Assessment
RID (rail)	Règlement Concernant le Transport International Ferroviaire des Marchandises Dangereuses – Regulation Concerning the Transport of Dangerous Goods by International Railway
RRR	Relative Risk Ranking
SHHSV	Stationary High-pressure Hydrogen Storage Vessel
SRB	Solid Rocket Boosters
SST	Shear Stress Transport Models
STP	Standard Temperature and Pressure
SUV	Suburban Utility Vehicle
SwRI	Southwest Research Institute
TNT	Trinitrotoluene
TPED	Transportable Pressure Equipment Directive
TRD	Thermal Relief Device
UFL	Upper Flammability Limit
UDL	Upper Detonability Limit
UN ECE	United Nations' Economic Commission for Europe
UVCE	Unconfined Vapor Cloud Explosion
WI	What-If Analysis

1

Introduction

In examining the history of hydrogen usage, it is perhaps informative to turn to the development of another energy resource, petroleum. Edwin Drake drilled the first oil well in Pennsylvania in 1859 to use extracted oil as a substitute for whale oil, the main lighting source and feedstock for consumer and chemical products at that time. That resource was hazardous to obtain and dwindling because of heavy exploitation, as is currently the case with oil. Petroleum presented many advantages over whale oil and solved a great many ecological and resource security problems associated with the old resource. Nevertheless, after a century and a half of use, petroleum has created new problems related to environmental pollution and energy security [1]. To address these problems, efforts have been dedicated to the gradual substitution of all ordinary fossil fuels (oil, coal, and natural gas) with new energy vectors, primarily because of the exhaustion of natural resources and the simultaneous rapid increase of energy demands worldwide.

Hydrogen is considered one of the most promising fuels for generalized use in the future, mainly because it is an energy-efficient, low-polluting, and renewable fuel. Hydrogen is versatile and clean, and with environmental benefits in mind, the production of hydrogen from renewable energy sources such as biomass, wind, solar, and nuclear sources is under consideration [2-3]. Yet, discussion of potential problems that widespread use of hydrogen might bring, as with whale oil and petroleum before it, is lacking. The identification and control of potential adverse consequences is an ethical requirement whenever any new technology is introduced, and certainly in the case of hydrogen, with such world-influencing potential [1].

The strategic areas of research under investigation and financed by research organizations worldwide with regard to hydrogen are the following:

- Clean hydrogen production from existing and novel processes.
- Storage, including hybrid storage systems.
- Basic materials, including those for electrolysers, fuel cells, and storage systems.
- Safety and regulatory issues required for the preparation of regulations and safety standards at a global level.
- Societal issues, including public awareness and preparations for the transition to a hydrogen energy economy.

Hydrogen could be an effective energy carrier if it can be rendered economically competitive and connected with an infrastructure providing a safe and environmentally acceptable energy system throughout the whole production, distribution, and end-use chain. Hydrogen has been, for more than a century, produced and used with a high safety record for commercial and industrial purposes, such as refinery and chemical processes and rocket propulsion, or as a radiolytically produced byproduct in nuclear boiling water reactors. Industry has significant experience in the safe handling of hazardous materials in chemical plants, where only well-trained personnel come into contact with hydrogen. Nevertheless, the wide use of hydrogen as an energy carrier will result in its use by laypersons, necessitating different safety regulations and technologies, which are now under development [4-5].

One of the major issues affecting the acceptance of hydrogen for public use is the safety of hydrogen installations (production and storage units), as well as its applications (i.e., as vehicle fuel or home use). The hazards associated with the use of hydrogen can be characterized as physiological (frostbite and asphyxiation), physical (component failures and embrittlement), and chemical (burning or explosion), the primary hazard being inadvertently producing a flammable or explosive mixture with air [6-7]. As far as European countries are concerned, hazardous chemicals installations come under the so-called SEVESO II Directive (96/82/EC) for the control of major-accident hazards involving dangerous substances. Hydrogen is included in this directive, indeed with a stricter minimal quantity for the directive's implementation than any other ordinary fuel [8-9].

In recent works, theoretical [10] and computational [11] safety comparisons between hydrogen and other fuels do not allow a clear point of view for the safest one to be concluded. Indeed, in the past, there were circumstances in which hydrogen applications gave rise to severe accidents with significant economic and societal cost, affirming the need for augmented safety measures wherever hydrogen is handled [12]. Undoubtedly, the need for safety measures should be pointed out when loss prevention and public safety are concerned. This supposes the knowledge of potential hazards plus the determination of risk zones around installations handling hydrogen.

As noted by Guy [13] in his article "The Hydrogen Economy," hydrogen is largely produced as synthesis gas for use in chemical production (e.g., ammonia and methanol) or recovered as a byproduct for use in oil refineries. He further comments that while the safe handling of hydrogen by industry (especially industrial gas companies) is well understood, use of hydrogen in the public realm can be problematic. This book demonstrates that the need for safer production, storage, distribution, and use of hydrogen in all application sectors must indeed be well understood and acted upon if the envisaged hydrogen economy is to materialize and endure.

The remainder of the book is organized in the following manner. Chapter 2 provides an overview of hydrogen accidents over the years. This is followed in Chapter 3 by a discussion of the various properties of hydrogen associated with hazards in the gas and liquid phases. The manifestation of these properties as physiological, physical, and chemical hazards is discussed in Chapter 4. The hazards and ensuing risks associated with hydrogen storage facilities and hydrogen use as a transportation fuel are addressed in Chapters 5 and 6, respectively.

Chapter 7 deals with the application of the principles of inherently safer design [14] to the field of hydrogen safety. Enhancements in the safer handling and use of hydrogen by means of an appropriate safety management system are then covered in Chapter 8. Chapter 9 describes a unique effort aimed at facilitating the safe introduction of hydrogen as an energy carrier and removing safety-related obstacles—the EC (European Commission) Network of Excellence, known as HySafe [15]. This is followed in Chapter 10 by an exposition of the value of studying lessons learned from case histories in making safety improvements in all areas, specifically here with respect to hydrogen usage.

Chapter 11 deals with the important issue of the effects of hydrogen on materials of construction; the subject of Chapter 12 is future needs in the area of hydrogen safety. Finally, Chapter 13 provides an overview of various guidelines and procedural approaches to hydrogen safety, and Chapter 14 gives concluding remarks.

The material has been arranged in a sequence that is intended to integrate the traditional view of hydrogen safety—hazardous properties and industrial risks in storage and use—with the equally important, yet often overlooked, safety aspects associated with the management and social sciences (for example, safety management systems and safety culture). It is therefore hoped that this book will be helpful to practitioners and researchers in a variety of fields and with a range of backgrounds and experience.

References

1. Cherry, R.S., A hydrogen utopia? *International Journal of Hydrogen Energy*, 29, 125, 2004.
2. Akansu, S.O., Dulger, Z., Kahraman, N., and Veziroglu, T.N., Internal combustion engines fueled by natural gas: hydrogen mixtures, *International Journal of Hydrogen Energy*, 29, 1527, 2004.
3. Rigas, F., and Sklavounos, S., Evaluation of hazards associated with hydrogen storage facilities, *International Journal of Hydrogen Energy*, 30, 1501, 2005.
4. Momirlan, M., and Veziroglu, T.N., Current status of hydrogen energy, *Renewable and Sustainable Energy Reviews*, 6, 141, 2002.
5. EUR 22002, *Introducing Hydrogen as an Energy Carrier*, European Commission, Directorate-General for Research Sustainable Energy Systems, 2006.
6. Schulte, I., Hart, D., and van der Vorst, R., Issues affecting the acceptance of hydrogen fuel, *International Journal of Hydrogen Energy*, 29, 677, 2004.

7. Dincer, I., Technical, environmental and exergetic aspects of hydrogen energy systems, *International Journal of Hydrogen Energy*, 27, 265, 2002.
8. European Economic Community, *On the Control of Major-Accident Hazards Involving Dangerous Substances*, Directive 96/82/EC, Brussels, 1996.
9. Kirchsteiger, C., Availability of community level information on industrial risks in the EU. *Process Safety and Environmental Protection*, 78, 81, 2000.
10. Institute of Chemical Engineers, *Accident Database* (CD form), Loughborough, U.K., 1997.
11. Taylor, J.R., *Risk Analysis for Process Plants, Pipelines and Transport*, Chapman & Hall, London, 1994, 102.
12. Center for Chemical Process Safety, *Guidelines for Hazard Evaluation Procedures*, American Institute of Chemical Engineers, New York, 1992, 69.
13. Guy, K.W.A., The hydrogen economy, *Process Safety and Environmental Protection*, 78 (4), 324–327, 2000.
14. Kletz, T., and Amyotte, P., *Process Plants: A Handbook for Inherently Safer Design*, CRC Press/Taylor & Francis Group, Boca Raton, FL, 2010.
15. Jordan, T., Adams, P., Azkarate, I., Baraldi, D., Barthelemy, H., Bauwens, L., Bengaouer, A., Brennan, S., Carcassi, M., Dahoe, A., Eisenrich, N., Engebo, A., Funnemark, E., Gallego, E., Gavrikov, A., Haland, E., Hansen, A.M., Haugom, G.P., Hawksworth, S., Jedicke, O., Kessler, A., Kotchourko, A., Kumar, S., Langer, G., Stefan, L., Lelyakin, A., Makarov, D., Marangon, A., Markert, F., Middha, P., Molkov, V., Nilsen, S., Papanikolaou, E., Perrette, L., Reinecke, E.-A., Schmidtchen, U., Serre-Combe, P., Stocklin, M., Sully, A., Teodorczyk, A., Tigreat, D., Venetsanos, A., Verfondern, K., Versloot, N., Vetere, A., Wilms, M., and Zaretskiy, N., Achievements of the EC Network of Excellence HySafe, *International Journal of Hydrogen Energy*, 36 (3), 2656–2665, 2011.

2

Historical Survey of Hydrogen Accidents

A historical survey reveals that there have been quite a few severe accidents in the industrial and transport sectors because of hydrogen use. The determinant causes for a major event may be classified in one of the following categories [1]:

- Mechanical or material failure
- Corrosion attack
- Overpressurization
- Enhanced embrittlement of storage tanks at low temperatures
- Boiling liquid expanding vapor explosion
- Rupture due to impact by shock waves and missiles from adjacent explosions
- Human error

Well-known accident databases, such as those of the United Nations Environment Programme and the Organization for Economic Co-operation and Development, MHIDAS (Major Hazardous Incident Data Service), and BARPI (Bureau d'Analyse des Risques et Pollutions Industriels), have been used to collect and classify past accidents related to hydrogen applications. The results are shown in Table 2.1 [2-8].

Although this is not a complete list of all accidents involving hydrogen, Table 2.1 gives a clear idea of the potential hazards stemming from careless use, storage, or transmission of hydrogen. This is why such databases should be continuously updated. These characteristic examples prove that, in practice, hydrogen is liable to cause major chemical accidents posing a considerable risk not only for onsite but also for offsite damage.

Typical examples of such accidents with consequences on people and property are quoted below and may be found in detail in the relevant literature and databases [7-10].

TABLE 2.1

Summary of Major Accidents Related to Hydrogen

Year/Date	Location	Origin of Accident/ Activity	Death	Injury	Evacuated
2001/05.01	Oklahoma	Fire/transport (trailer)	1	1	15
2001/04.18	Labadie, Montana	Fire/power plant	NA	NA	NA
2000/09.03	Gonfreville-L'Orcher, Havre, France	Explosion/ chemical plant	-	12	-
2000/02.10	Tomakomai, Hokkaido, Japan	Fire/refinery	-	-	-
1999/05.07	Panipat, India	Fire/refinery	5	-	-
1999/04.08	Hillsborough, Tampa, Florida	Fire and explosion/ power station	3	50	38
1998/09.15	Torch, Canada	Fire/nuclear industry	NA	NA	NA
1998/06.08	Auzouer en Touraine, France	Explosion and fire/fine chemicals	-	1	200
1993/NA	Krasnaya Tura, Russia	Cloud explosion/ pipeline	-	4	-
1992/04.22	Jarrie, Isere, France	Fire/chemical plant	1	2	-
1992/01.18	Pennsylvania	Fire/NA	1	3	-
1992/10.16	Sodegaura, Japan	Explosion/ refinery	10	7	-
1992/08.08	Wilmington, Delaware	Explosion/ refinery	-	16	-
1992/08.28	Hong Kong	Explosion/ power plant	2	19	-
1991/05.16	Kakuda, Miyagi, Japan	Explosion/space rocket facilities	-	-	-
1991/02.14	Daesan, Korea	Explosion/ petrochemicals	-	2	-
1991/10.-	Hanau, Frankfurt, Germany	Explosion/ optical fiber production	NA	NA	NA
1990/07.25	Birmingham, UK	Fire and gas cloud	-	>60	70,050
1990/NA	Czechoslovakia	Explosion	15	26	NA
1990/04.29	Ottmarsheim, France	Fire	NA	NA	NA
1989/NA	USA	Jet fire/pipeline	7	8	-
1989/10.23	Pasadena, Texas	Explosion/ chemical plant	23	314	NA

TABLE 2.1 (*Continued*)

Summary of Major Accidents Related to Hydrogen

Year/Date	Location	Origin of Accident/ Activity	Death	Injury	Evacuated
1988/06.15	Genoa, Italy	Explosion	3	2	15,000
1986/01.28	Kennedy Space Center, Florida	*Challenger* explosion/ space center	7	119	-
1977/NA	Gujarat, India	Explosion/ chemical plant	5	35	NA
1975/04.05	Ilford, Essex, UK	Explosion/ chemical plant	1	-	-
1972/NA	Netherlands	Explosion	4	40	NA
1937/05.06	Hindenburg, Lakehurst (Figure 2.1)	Fire	36	NA	NA

Note: NA, not available; -, none.

2.1 Significant Disasters

2.1.1 The Hindenburg Disaster

The largest aircraft ever built were the Zeppelins LZ-129 *Hindenburg* and her sister ship LZ-130 *Graf Zeppelin II.* They were constructed by Luftschiffbau Zeppelin in 1935. The *Hindenburg,* named after the president of Germany, Paul von Hindenburg, constructed with an aluminum framework, was 245 m long, 41 m in diameter, containing 211,890 m^3 of gas in 16 cells, with a useful lift of 112,000 kg, powered by four 820 kW engines giving it a maximum speed of 135 km/h. It was covered with cotton fabric impregnated with cellulose acetate butyrate resin containing iron oxide and aluminum powder. The *Hindenburg* made its first flight in March 1936 and was filled with the highly flammable hydrogen as the buoyancy gas, instead of the noncombustible helium, due to a United States military embargo on helium. Aiming at the prevention of any possible fire on the aircraft from hydrogen leaks, the German engineers used various safety measures, among which was the above special coating of the fabric to render it electrically conductive to avoid sparks from static electricity [11].

The disaster occurred at Lakehurst, New Jersey, on May 6, 1937 (Figure 2.1) and was for many years under investigation to identify the reasons that caused the ignition of hydrogen. It is remarkable that until that moment, Zeppelins had an impressive safety record, with no passenger injured on any of these airships, which were widely seen as symbols of German and Nazi superiority. Moreover, the *Graf Zeppelin* had flown safely for more than 1.6 million km (1 million miles) including the first complete circumnavigation

FIGURE 2.1
Hindenburg burning. (From Google Images.)

of the globe. Yet, in this accident, which terminated the use of airships as passenger carriers, the ignition of hydrogen proceeded rapidly to fire toward the tail section of the craft, completely destroying it. The fire was almost simultaneously succeeded by an explosion that quickly engulfed the 240-ton craft, causing it to crash onto the ground, killing 36 people. Thorough investigation of the accident showed that the outer shell and the paint of the airship, used only on the *Hindenburg*, were flammable and could be eventually ignited by electrical sparks. Cellulose acetate butyrate is flammable by itself, but its flammability is considerably increased by the addition of iron oxide and aluminum powder. In fact, iron oxide and aluminum are sometimes used as components of solid rocket propellants or thermite.

In addition, the airship's outer shell was separated from the aluminum frame by nonconductive ramie cords, thus permitting static electricity accumulation on the shell. Prevailing atmospheric conditions at the time the accident occurred could generate considerable electrostatic discharge activity on the airship. Furthermore, the mooring lines were wet (i.e., conductive) and connected to the aluminum frame. When the wet mooring lines touched the ground, they grounded the aluminum frame, resulting in the formation of an electrical discharge from the electrically charged shell to the grounded frame [11].

Although the airship was completely destroyed when hydrogen caught fire, the triggering cause was the use of hazardous materials for the outer shell and bad design with regard to permitting static electricity development on it. Since no death was directly related to the hydrogen burning, probably the airship would have been destroyed even if helium was used instead of hydrogen due to the shell burning and the loss of buoyancy. Nevertheless, taking into account that hydrogen is extremely flammable, nowadays, helium is used exclusively instead of hydrogen in buoyancy-driven airships, thus resulting in inherently safer systems (as discussed in Chapter 7).

2.1.2 Hydrogen Release during Maintenance Work

On October 23, 1989, at Pasadena, Texas, a massive and destructive hydrogen vapor cloud explosion occurred in a polyethylene plant, causing loss of life to 23 people, injuries to 314 others, and extensive damage to the whole plant. The explosion strength was estimated to be equivalent to 2.4 tons of the explosive TNT. The accident resulted from a release of about 40,000 kg of process gas containing ethylene, isobutene, hexane, and hydrogen during maintenance on a reaction loop line. The U.S. Occupational Safety and Health Administration (OSHA) detailed numerous defects in the management of the installation, which in brief were the following:

- Lack of a hazard assessment study (as discussed in Chapter 8).
- Inadequate safety distances between the control room and the reactor that did not allow emergency shutdown actions and safe evacuation of the personnel during the initial vapor release.
- Lack of an effective permit system for the control of maintenance activities (as discussed in Chapter 8).
- Erroneous design of building ventilation intakes, so that in the case of gas release from the process unit, the escaping gas could be trapped in adjacent buildings.

2.1.3 Pressurized Hydrogen Tank Rupture

In 1991 at Hanau, Frankfurt, Germany, a tank containing 100 m^3 of hydrogen pressurized at 45 bar burst without apparent reason, while it was stored outdoors in an industrial plant. The shock wave and the missiles of the tank shell caused heavy damage to the other units of the plant [12]. Investigations showed that welding in the metallic shell suffered from extensive cracks from the inner side to the outside. Likely, the corners along the welding caused concentration of stresses, so that the first cracks appeared in those parts. Under the influence of hydrogen, cracks grew much faster than normal, leading to material failure and subsequent mechanical explosion. As a result, all comparable tanks in the country were checked, manufacturing rules were revised providing upper limits for corners, and new test methods for detecting cracks in early stages during operation were established. Undoubtedly, this accident contributed to improvements in hydrogen tank safety.

2.1.4 The *Challenger* Disaster

The first launch of the space shuttle *Challenger* was on April 12, 1981, after seven years of development (1972–1979). Its three main engines from Rocketdyne operated with liquid hydrogen and liquid oxygen as fuel and oxidizer, respectively. The external tank from Martin Marietta contained 616 ton liquid oxygen (1991 L) and 102 ton liquid hydrogen (14,500 L). The solid rocket boosters (SRB) from Thiokol each contained 503 ton of fuel (16 percent atomized aluminum powder as fuel, 69.83 percent ammonium perchlorate as oxidizer, 0.17 percent iron powder as catalyst, 12 percent polybutadiene acrylic acid acrylonite as binder, and 2 percent epoxy curing agent).

On January 28, 1986, at 11:38 a.m. Eastern Standard Time, the *Challenger* space shuttle left Pad 39B at the Kennedy Space Center in Florida for Mission 51-L. This was the tenth flight of Orbiter *Challenger*. Seventy-three seconds later the space shuttle was completely destroyed owing to an explosion of the hydrogen tank (Figure 2.2). All seven crew members (six astronauts and one civilian) were killed.

The cause of the explosion was determined to be an O-ring failure in the right SRB. The unusually cold weather (see Figure 2.2), beyond the tolerances for which the rubber seals were approved, most likely caused the O-ring failure. The temperature at ground level at Pad 39B was 36°F (2.2°C); this was 15°F (8.3°C) colder than any other previous launch by NASA [11].

FIGURE 2.2 (See color insert.)
Space Shuttle *Challenger* at lift-off, its smoke plume after in-flight breakup, and icicles that draped the Kennedy Space Center on that day. (From Great Images in NASA.)

2.2 Incident Reporting

Reporting of incidents related to hydrogen and analyzing the principal causes are useful for sharing lessons learned with the private and public sectors. For this purpose the H_2Incidents Database has been created by the Pacific Northwest National Laboratory with funding from the U.S. Department of Energy (available at http://www.h2incidents.org/). In this database, incidents and near-misses are reported without including the names of the companies and other details; this confidentiality encourages reporting of events. The incidents are classified according to settings, equipment, damage and injuries, probable causes, and contributing factors.

The most important records from this database in percentages of incidents by the end of the year 2010 are shown in Figures 2.3 to 2.6. Thus, the percentage of hydrogen incidents in various settings as reported in the H_2Incidents Database for a total of 209 incidents is depicted in Figure 2.3. This figure shows that laboratory incidents are by far the most frequent nowadays, but this is expected to change in the years to come as there is a movement from the intense hydrogen research of today to the more widespread utilization of hydrogen.

The percentage of types of equipment involved in these incidents is shown in Figure 2.4 for a total of 334 incidents. As is the case for technological incidents in general, the simplest equipment (e.g., piping, fittings, valves, and storage tanks) is most frequently involved in an incident because less attention is usually paid to them.

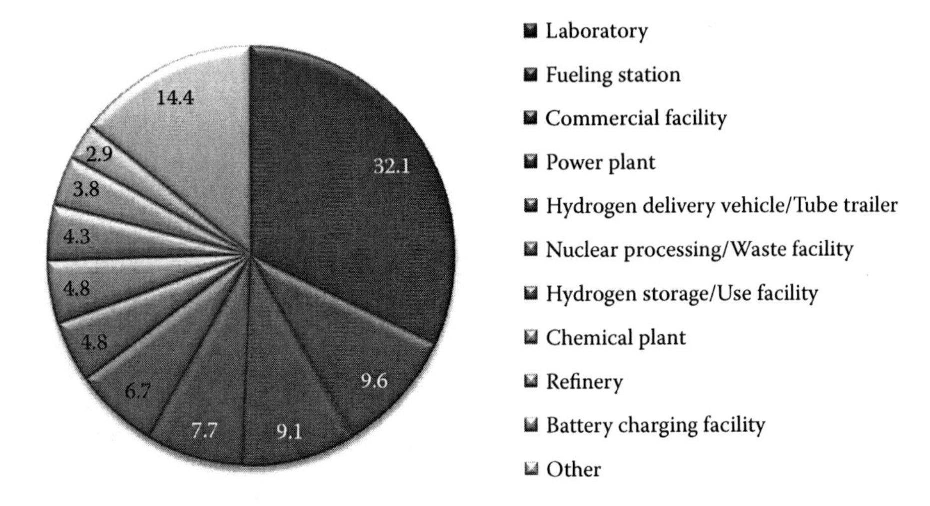

FIGURE 2.3
Percentage of incidents related to hydrogen in various settings, as reported in the H_2Incidents Database.

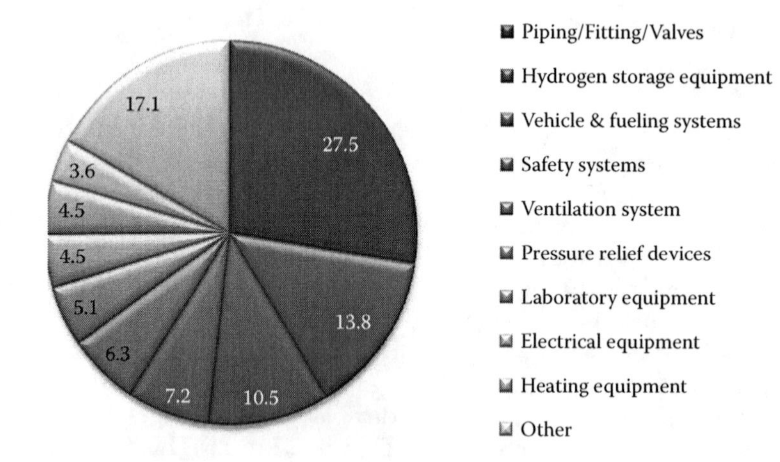

FIGURE 2.4
Percentage of different types of equipment involved in the incidents reported in the H_2Incidents Database.

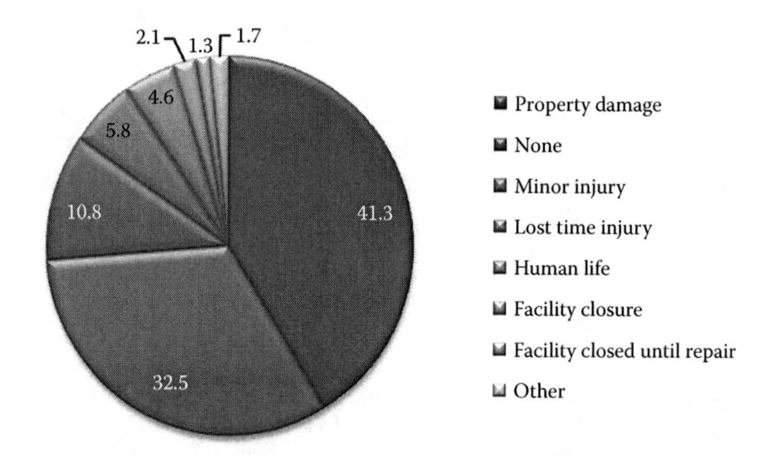

FIGURE 2.5
Percentage of types of damage and injuries from hydrogen incidents as reported in the H_2Incidents Database.

It is clear from Figure 2.5, in which percentages of damage and injuries during hydrogen incidents are shown, that for a total of 240 incidents only a small proportion results in loss of human life (4.6 percent). This is because special mitigation measures are usually taken, given knowledge of the severity of such incidents.

With regard to probable causes of hydrogen incidents, Figure 2.6 verifies for 291 incidents a situation that is well known in the process industries. Indeed, although equipment failure is the most common cause of incidents, in fact human error is hidden inside almost every other cause,

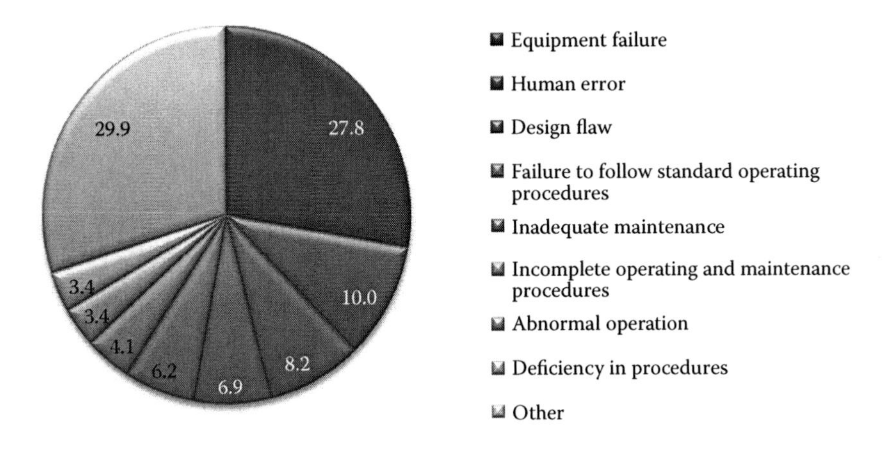

FIGURE 2.6
Percentage of probable causes for hydrogen incidents as reported in the H_2Incidents Database.

including equipment failure, design flaw, and failure to follow standard operating procedures.

The lessons learned from this analysis are the following:

- Many incidents occur nowadays in laboratories because of the intensified research on production, storage, and utilization of hydrogen.
- Most incidents occur in the simplest equipment, such as piping, fittings, and valves, to which one does not usually pay enough attention.
- Most incidents result in property damage, but only a few in injuries, probably due to the limited quantities of hydrogen still used and the mitigation measures taken.
- The most probable cause is equipment failure, while lack of situational awareness and human error are the most common contributing factors.

The most recent incidents due to hydrogen that were recorded in 2010, according to the H_2Incidents Database, are shown in Table 2.2. This cause and consequences analysis gives an idea of the broad range of factors that can lead to an accident when using, storing, or transporting hydrogen. Figure 2.7 shows one example from the database.

In addition to the H_2Incidents Database, many other remarkable efforts have been developed or are under way, aimed at collecting and offering valuable information on past accidents to assist in the composition of new safety regulations, codes of practice and standards, and in the prevention of similar accidents by supplying useful data for both qualitative and quantitative risk assessment (QRA).

Among them, the network of excellence project *HySafe* (Safety of Hydrogen as an Energy Carrier) greatly contributes to the successful transition of

TABLE 2.2

Most Recent Accidents (2010) due to Hydrogen That Were Reported to the H_2*Incidents Database* (Intended for Public Use)

Incident Type and Description	Damage and Injuries	Cause and Ignition	Lessons Learned/Suggestions/ Mitigation Measures
Hydrogen gas explosion in municipal refuse incineration facility. The workers injected water to remove some blockage in clinker, and the water reacted with incinerated aluminum ash to form hydrogen. *Date reported:* 11/16/2010	Three workers were seriously burned by high-temperature gas that spouted from the inspection door, and one of them died 10 days later. Property damage.	The workers injected water to remove some blockage in clinker, and the water reacted with incinerated aluminum ash to form hydrogen, which caused the explosion. *Probable causes:* Human error and deficiency in procedures. *Ignition source:* Either hot clinker or sparks from chisel.	Aluminum should have been separated from the refuse prior to feeding it to the incinerator.
Hydrogen explosion and fire in a styrene plant in a large petrochemical complex. The explosion followed the release of about 30 kilograms of 700-psig hydrogen gas from a burst flange into a compressor shed. *Date reported:* 11/12/2010	Two men were killed and two others injured. Property damage.	The operators were bringing the plant online increasing the hydrogen circulation pressure. *Probable causes:* Equipment failure, inadequate equipment, and hazard not identified in advance. *Ignition source:* Unknown.	Hydrogen gas was released through a failed 19-inch diameter gasket and ignited under the roof of the compressor shed where it was partially confined. Areas where hydrogen may be released should be not be confined with walls and roofs.

Fatal accident due to bacterial hydrogen production in storage tank. The accident took place at an onshore processing facility for slop water from offshore petroleum industry. *Date reported:* 5/03/2010	Two operators were trying to remove the lid from a manhole on top of a 1600-cubic meter storage tank. To open the rusted bolts holding the lid in place, they used a cutting disk. One man was thrown over from the explosion and killed. Property damage.	The hydrogen inside the tank from biological fermentation exploded because of sparks from the cutting disk. *Probable causes:* Human error, flammable mixture in confined area, and hazard not identified in advance. *Ignition source:* Sparks from angle grinder.	Avoid anaerobic and stagnant conditions in slop water by providing sufficient circulation and/or aeration. Use inert atmosphere or open-top floating roof tank to avoid space where an explosive mixture could form. Use nonsparking tools where explosive atmospheres may develop.
Pickup collision with tractor pulling hydrogen tube trailer (Figure 2.7). The collision caused damage to tubes, valves, piping, and fittings, resulting in hydrogen release and fire. The hydrogen ignited and burned the rear of the semitrailer. *Date reported:* 1/21/2010	The tractor-semitrailer driver was killed as a result of blunt force trauma. The driver of the pickup truck received non-life-threatening injuries. Damage, cleanup, and lost revenues were estimated at $155,000.	*Probable cause:* Vehicle collision. *Ignition source:* No ignition source defined.	Increase physical protection, shielding, and securing of transported hydrogen tubes.
Refinery hydrocracker pipe rupture and release of explosive mixture with hydrogen. A refinery hydrocracker effluent pipe section ruptured and released a mixture of gases, including hydrogen, which instantly ignited on contact with the air, causing an explosion and a fire. *Date reported:* 1/12/2010	An operator who was checking a field temperature panel at the base of the reactor and trying to diagnose the high-temperature problem was killed. A total of 46 other plant personnel were injured. Property damage.	Excessive high temperature, likely in excess of 760°C, initiated in one of the reactor beds spread to adjacent beds and raised the temperature and pressure of the effluent piping to the point where it failed. *Probable causes:* Failure to follow standard operating procedures, human error, and incorrect engineering hazard analysis calculation. *Ignition source:* No ignition source defined.	Management must provide an operating environment conducive for operators to follow emergency shutdown procedures when required. Some backup system of temperature indicators should be used so that the reactors can be operated safely in case of instrument malfunction. Process hazard analysis must be based on actual equipment and operating conditions.

FIGURE 2.7 (See color insert.)
Scene of a hydrogen tube trailer accident (From "H2Incidents Database" intended for public use, http://www.h2incidents.org/docs/265_1.pdf)

Europe to sustainable development based on hydrogen use. One of the work packages of *HySafe* under construction is the European Hydrogen Incident and Accident Database (HIAD), dedicated to offer a collaborative and communicative database in the form of an open web-based information system [13]. Because of its significance, the *HySafe* project is presented in a separate chapter in this book (Chapter 9).

References

1. Federal Institute for Materials Research and Testing, *Hydrogen Safety*, German Hydrogen Association, Brussels, 18–21, 2002.
2. Rigas, F., and Sklavounos, S., Hydrogen safety, in *Hydrogen Fuel: Production, Transport and Storage*, R. Gupta (Ed.), CRC Press/Taylor & Francis, Boca Raton, FL, 2008, 537–538.
3. Rigas, F., and Sklavounos, S., Evaluation of hazards associated with hydrogen storage facilities, *International Journal of Hydrogen Energy*, 30, 1501–1510, 2005.
4. Institution of Chemical Engineers (IChemE), *Accident Database* (CD form), Loughborough, U.K., 1997.

5. Rosyid, O.A. System-Analytic Safety Evaluation of the Hydrogen Cycle for Energetic Utilization, Dissertation, Otto-von-Guericke University, Magdeburg, Germany, 2006.
6. Khan, F.I., and Abbasi, S.A., Major accidents in process industries and an analysis of causes and consequences, *Journal of Loss Prevention in the Process Industries*, 12, 361–378, 1999.
7. Bethea, R.M. *Explosion and Fire at Pasadena, Texas,* American Institute of Chemical Engineers, New York, 1996.
8. Ministère chargé de l'environnement, France, *Explosion dans une unité de craquage d'une raffinerie*, No 19423, 2000.
9. Center for Chemical Process Safety. *Guidelines for Hazard Evaluation Procedures*, American Institute of Chemical Engineers, New York, 1992, 69.
10. Warren, P. *Hazardous Gases and Fumes*, Butterworth-Heinemann, Oxford, 1997, 96.
11. Zuettel, A., Borgschulte, A., and Schlapbach, L. (Eds.), *Hydrogen as a Future Energy Carrier*, Wiley-VCH Verlag, Berlin, Germany, 2008, 14–20.
12. Hord, J. Is hydrogen a safe fuel? *International Journal of Hydrogen Energy*, 3, 157, 1978.
13. Kirchsteiger, C., Vetere Arellano, A.L., and Funnemark, E., Towards establishing an international hydrogen incidents and accidents database (HIAD) *Journal of Loss Prevention in the Process Industries*, 20, 98, 2007.

3

Hydrogen Properties Associated with Hazards

3.1 General Consideration

With regard to the relative rotation of the nuclear spin of the individual atoms, the following forms of hydrogen are defined: *o*-H_2, *ortho*-hydrogen; *p*-H_2, *para*-hydrogen; *e*-H_2, equilibrium hydrogen; and *n*-H_2, normal hydrogen. The form *e*-H_2 corresponds to the equilibrium concentration at definite temperatures.

Normal hydrogen is a mixture of 75 percent *ortho*-hydrogen and 25 percent *para*-hydrogen in thermodynamic equilibrium at room temperature (293.15 K or 20°C). These two forms are distinguished by the relative rotation of the nuclear spin of the individual atoms in the molecule. In *ortho*-hydrogen the molecules spin in the same direction (symmetric with parallel nuclear spins, ↑↑), whereas in *para*-hydrogen they spin in the opposite direction (antisymmetric with antiparallel nuclear spins, ↑↓) [1].

During the process of cooling of hydrogen from room temperature (RT) to its normal boiling point (NBP = 21.2 K) the equilibrium concentration of *ortho*-hydrogen drops from 75 percent at RT to 50 percent at 77 K and finally to 0.2 percent at NBP. The noncatalyzed conversion rate is very slow, with a half-life of the conversion greater than one year at 77 K. The conversion reaction from *ortho*- to *para*-hydrogen is exothermic and the heat of conversion is temperature dependent [2]. The heat of conversion is 670 J/g for 75 percent *o*-H_2 which is much higher than the heat of evaporation, 447 J/g. As a result of this internal heating, the loss of hydrogen from a storage vessel is enormous and reaches 30 percent after only 48 h [3]. Thus, in LH_2 (liquefied hydrogen) production facilities catalysts should be employed to accelerate this conversion. The transformation from *ortho*- to *para*-hydrogen can be catalyzed by a number of surface-active and paramagnetic materials. For instance, *n*-H_2 can be adsorbed on charcoal cooled with liquid hydrogen and desorbed in the equilibrium mixture (*e*-H_2). The conversion may take only a few minutes, if a highly active form of charcoal is used. Other suitable ortho–para catalysts are metals such as tungsten, nickel, or any paramagnetic oxides like chromium or gadolinium oxides. The nuclear spin is reversed without breaking the H-H bond [1, 3].

The physical properties of these two molecular forms differ slightly, but their chemical properties are practically identical, thus presenting the same chemical behavior with regard to chemical hazards [4].

Table 3.1 gives selected thermophysical, chemical, and combustion properties of hydrogen [1, 4-12]. Some of the key overall properties of hydrogen that are relevant to its employment as an engine fuel are also listed in this table together with the corresponding values of methane, the other promising gaseous fuel for engine applications, and those of gasoline [6]. Where property values of gasoline could not be found, these are represented by iso-octane vapor [7] or the arithmetic average of normal heptane and octane [8].

3.2 Hydrogen Gas Properties Related to Hazards

Detection: In atmospheric conditions, hydrogen gas is colorless, odorless, and not detectable in any concentration by human senses. It is not toxic but can cause asphyxiation by diluting the oxygen in the air, with a limiting oxygen concentration equal to 19.5 percent by volume, under which the atmosphere is considered oxygen deficient. Furthermore, the inability to detect hydrogen renders it a latent fuel, ready to ignite.

Volumetric leakage: With the anticipated large-scale introduction of hydrogen as an energy carrier, its volumetric leakage from containers and pipelines is expected to be 1.3–2.8 times as large as gaseous methane leakage and approximately 4 times that of air under the same conditions. Thus comes the rule: "airproof is not hydrogen-proof." On the other hand, any released hydrogen has the potential to disperse rapidly by fast diffusion, turbulent convection, and buoyancy, thus considerably limiting its presence in the hazardous zone [1].

Buoyancy: As shown in Table 3.1, hydrogen gas is about 14 times lighter than air in normal conditions (NTP) and this is why any leak quickly moves upward, thus reducing ignition hazards. Nevertheless, saturated vapor is heavier than air and will remain close to the ground until the temperature rises. Buoyant velocities range from 1.2 to 9 m/s in NTP air since they depend on the difference of air and vapor densities. Thus, the cold dense fuel vapors produced by LH_2 spills will initially remain close to the ground and then rise more slowly than standard temperature and pressure fuel gases [4, 13].

TABLE 3.1

Selected Comparative Properties of Hydrogen, Methane, and Gasoline

	Value			
Property	**Hydrogen**	**Methane**	**Gasoline**	**References**
Molecular weight	2.016	16.043	~107.0	6,8,11
Normal melting point (K)	14.1	90.68	213	11,12
Normal boiling point (K)	20.268	111.632	310–478	4,6,8,11,12
Critical temperature (K)	32.97–33.1	190	—	11,12
Critical pressure (MPa or atm)	1.8 or 12.8	4.6	—	11,12
Density of vapor at NBP (kg/m³)	1.338	73.4	—	1,4,12
Density of liquid at NBP (kg/m³)	70.78	423.8	745 (at STP)	1,4,12
Density of gas at NTP (g/m³)	82 (at 300 K) 83.764	717 651.19	5,110 ~4,400	7,8
Density of gas at STP (g/m³)	84 89.87	650 657 (at 298.2 K)	4,400	1 11, 12
Heat of fusion at 14.1 K (kJ/kg)	58	0.94 (kJ/mol)	—	11
Heat of vaporization (kJ/kg)	445.6 447	509.9	250–400	1,11
Heat of combustion (low) (kJ/g)	119.93 119.7	50.02 46.72	44.5 44.79	1,4,8 7
Heat of combustion (high) (kJ/g)	141.86 141.8 141.7	55.53 55.3 52.68	48 48.29 —	1,4,8 1 7
Flammability limits in NTP air (vol%)	4.0–75.0 —	5.3–15 —	1.0–7.6 1.2–6.0 1.4–7.6	1,4,6,8 7 12
Flammability limits in NTP oxygen (vol%)	4.1–94.0	—	—	4
Detonability limits in NTP air (vol%)	18.3–59.0 13.5–70	6.3–13.5 —	1.1–3.3 —	1,4,6,8 9
Detonability limits in NTP oxygen (vol%)	15–90	—	—	4
Stoichiometric composition in air (vol%)	29.53	9.48	1.76	1,4,8
Minimum ignition energy in air (mJ)	0.017 0.02 0.14	0.29 0.28 —	0.24 0.25 0.024	1,4,8 7 6
Autoignition temperature (K)	858	813	501-744 500-750	1,4,8 7
Adiabatic flame temperature in air (K)	2,318 —	2,148 2,190	~2,470 —	1,4,6-8 7
Thermal energy radiated from flame to surroundings (%)	17–25	23–33	30–42	4,6,8

—continued

TABLE 3.1 (*Continued*)

Selected Comparative Properties of Hydrogen, Methane, and Gasoline

	Value			
Property	**Hydrogen**	**Methane**	**Gasoline**	**References**
Burning velocity in NTP air (cm/s)	265–325	37–45	37–43	4,8
Burning velocity in STP air (cm/s)	346	45	176	1
Detonation velocity in NTP air (km/s)	1.48–2.15	1.39–1.64	1.4–1.7	4,8
Detonation velocity in STP air (km/s)	1.48–2.15	1.4–1.64	1.4–1.7	1
Energy of stoichiometric mixture in NTP air (MJ/m^3)	3.58	3.58	3.91	10
Velocity of sound of vapor (m/s)	305	—	—	4
Velocity of sound of liquid (m/s)	1273	—	—	4
Diffusion coefficient in NTP air (cm^2/s)	0.61	0.16	0.05	4,6,8
Diffusion coefficient in STP air (cm^2/s)	0.61	0.16	0.05	1
Buoyant velocity in NTP air (m/s)	1.2–9	0.8–6	Nonbuoyant	4,8
Limiting oxygen index (vol%)	5.0	12.1	11.6	1,8
Maximum experimental safe gap in NTP air (cm)	0.008	0.12	0.07	4,8
Quenching gap in NTP air (cm)	0.064	0.203	0.2	4,7,8
Detonation induction distance in NTP air	L/D ~100	—	—	8

Note: STP (standard temperature pressure): 273.15 K (0°C), 101.3 kPa (1 atm); NTP (normal temperature pressure): 293.15 K (20°C), 101.3 kPa; NBP (normal boiling point): boiling point at 101.3 kPa.

Flame visibility: A hydrogen-air-oxygen flame is nearly invisible in daylight irradiating mostly in the infrared and ultraviolet region. Thus, any visibility of a hydrogen flame is caused by impurities such as moisture or particles in the air. Hydrogen fires are readily visible in the dark and large hydrogen fires are detectable in daylight by the "heat ripples" and the thermal radiation to the skin [8]. At reduced pressures a pale blue or purple flame may be visible. An obvious hazard resulting from this property may be severe burns on persons exposed to hydrogen flames due to the ignition of hydrogen gas escaping from leaks.

Flame temperature: The flame temperature for 19.6 percent by volume hydrogen in air has been measured as 2318 K [1]. More information for deflagration and detonation temperatures and pressures are given in Chapter 4, Section 4.3.4.2, derived by the Gordon-McBride code [5, 14].

Burning velocity: Burning velocity in air is the subsonic velocity at which a flame of a flammable fuel-air mixture propagates. For hydrogen this velocity ranges from 2.65 to 3.46 m/s, depending on pressure, temperature, and mixture composition. This high burning velocity of hydrogen, which is one order of magnitude higher than that of methane (maximum burning velocity in air at STP: 0.45 m/s), indicates its high explosive potential and the difficulty of confining or arresting hydrogen flames and explosions [1, 4].

Thermal energy radiation from flame: Exposure to hydrogen fires can result in significant damage from thermal radiation, which depends largely on the amount of water vapor in the atmosphere. In fact, atmospheric moisture absorbs the thermal energy radiated from a fire and can reduce it considerably. The intensity of radiation from a hydrogen flame at a specific distance depends on the amount of water vapor present in the atmosphere and is expressed by the equation [4]:

$$I = I_o \cdot e^{-0.0046wr} \tag{3.1}$$

where

I_o = initial intensity (energy/time·area)
w = water vapor content (percent by weight)
r = distance (meters)

Limiting oxygen index: The limiting oxygen index is the minimum concentration of oxygen that will support flame propagation in a mixture of fuel vapors and air. For hydrogen, no flame propagation is observed at NTP conditions, if the mixture contains less than 5 percent by volume oxygen [4].

Joule–Thomson effect: When gases are expanded through a porous plug, or a small aperture or nozzle from high to low pressure, they usually are cooled. However, the temperature of some real gases increases when they are expanded at a temperature and pressure beyond the temperature and pressure conditions that define their Joule–Thompson (J-T) inversion curve. This maximum inversion temperature for hydrogen is 202 K at an absolute pressure of zero [1]. So, at temperatures and pressures greater than these, the temperature of hydrogen will increase upon expansion. With regard to safety, the increase of temperature as a result of the Joule-Thomson effect is not normally sufficient to ignite a hydrogen-air mixture. For instance, the temperature of hydrogen increases from 300 K to 346 K when it expands from a pressure of 100 MPa to 0.1 MPa. This increase in temperature is not sufficient to ignite hydrogen, whose autoignition temperature is 858 K at 1 atm and 620 K at low pressures [4].

3.3 Liquefied Hydrogen Properties Related to Hazards

All hazards accompanying hydrogen gas (GH_2) also exist with liquefied hydrogen (LH_2) due to its easy evaporation. Additional hazards should be taken into account when handling or storing liquid hydrogen because of that ease of evaporation.

Low boiling point: The boiling point of hydrogen at sea level pressure is 20.3 K. Any liquid hydrogen splashed on the skin or in the eyes can cause frostbite burns or hypothermia. Inhaling vapor or cold gas initially produces respiratory discomfort, and further breathing in can cause asphyxiation.

Ice formation: Vents and valves in storage vessels and dewars may be blocked by accumulation of ice formed from moisture in the air. Excessive pressure may then result in mechanical failure (container or component rupture), with jet release of hydrogen and potentially in a boiling liquid expanding vapor explosion (BLEVE).

Continuous evaporation: The storage of hydrogen as a liquid in a vessel results in continuous evaporation, changing its state to gaseous hydrogen. To equalize pressure, hydrogen gas must be vented to a safe location or temporarily collected safely. Storage vessels should be kept under positive pressure to prevent air from entering and producing flammable mixtures. Liquefied hydrogen may be contaminated with air condensed and solidified from the atmosphere or with trace air accumulated during liquefaction of hydrogen. The quantity of solidified air can increase during repeated refilling or pressurization of storage vessels, producing an explosive mixture with hydrogen.

Pressure rise: Liquefied hydrogen confined, for instance in a pipe between two valves, will eventually warm to ambient temperature, resulting in a significant pressure rise. Standard storage system designs usually assume a heat leak equivalent to 0.5 percent per day of the liquid contents. Considering liquefied hydrogen as an ideal gas, the pressure resulting from a trapped volume of liquefied hydrogen at one atmosphere vaporizing and being heated to 294 K is 85.8 MPa. However, the pressure is 172 MPa when hydrogen compressibility is considered [4].

High vapor density: The high density of the saturated vapor resulting immediately after release from a liquefied hydrogen storage vessel that is leaking causes the hydrogen cloud to move horizontally or downward for some time. This was shown experimentally by the National Aeronautics and Space Administration (NASA) in the Langley Research Center at the White Sands test facility in 1980 and simulated effectively later using a computational fluid dynamics (CFD) approach [13].

Electric charge buildup: Since electrical resistivity of liquefied hydrogen is about 10^{19} ohm-cm at 25 V, the electric current–carrying capacity is small and more or less independent of the imposed voltage. Investigation has shown that electric charge buildup in flowing liquefied hydrogen is not a great concern [4].

References

1. Zuettel, A., Borgschulte, A., and Schlapbach, L. (Eds.), *Hydrogen as a Future Energy Carrier*, Wiley-VCH Verlag, Berlin, Germany, 2008, Chap. 4.
2. Sullivan, N.S., Zhou, D., and Edwards, C.M., Precise and efficient *in situ* ortho–para-hydrogen converter, *Cryogenics*, 30, 734, 1990.
3. Yucel, S., Theory of ortho-para conversion in hydrogen adsorbed on metal and paramagnetic surfaces at low temperatures, *Physics Review B*, 39, 3104, 1989.
4. ANSI, *Guide to Safety of Hydrogen and Hydrogen Systems,* American Institute of Aeronautics and Astronautics, American National Standard ANSI/AIAA G-095-2004, Chap. 2.
5. Rigas, F., and Sklavounos, S., Hydrogen safety, in *Hydrogen Fuel: Production, Transport and Storage,* CRC Press/Taylor & Francis, Boca Raton, FL, 2008, Chap. 16.
6. Adamson, K.A., and Pearson, P., Hydrogen and methanol: a comparison of safety, economics, efficiencies and emissions, *Journal of Power Sources*, 86, 548, 2000.
7. Karim, A., Hydrogen as a spark ignition engine fuel, *International Journal of Hydrogen Energy*, 28, 569, 2003.
8. Hord, J., Is hydrogen a safe fuel? *International Journal of Hydrogen Energy*, 3, 157, 1978.
9. Baker, W.E., and Tang, M.J., *Gas, Dust and Hybrid Explosions*, Elsevier, Amsterdam, 1991, 42.

10. Rosyid, A., and Hauptmanns, U., System analysis: safety assessment of hydrogen cycle for energetic utilization, in *Proc. Int. Congr. Hydrogen Energy and Exhibition*, Istanbul, 2005.
11. Kirk-Othmer Encyclopedia of Chemical Technology, *Fundamentals and Use of Hydrogen as a Fuel,* 3rd ed., Vol. 4, Wiley, New York, 1992.
12. Wolfram Alpha, Computational Knowledge Engine, http://wolframalpha.com.
13. Rigas, F., and Sklavounos, S., Evaluation of hazards associated with hydrogen storage facilities, *International Journal of Hydrogen Energy*, 30, 1501–1510, 2005.
14. Gordon, S., and McBride, B.J., *Computer Program for Calculation of Complex Chemical Equilibrium Compositions and Applications*, NASA Reference Publication 1311, 1994.

4

Hydrogen Hazards

The hazards associated with hydrogen can be physiological (frostbite and asphyxiation), physical (component failures and embrittlement), or chemical (burning or explosion), the primary hazard being inadvertently producing a flammable or explosive mixture with air. Safety can be obtained only when designers and operational personnel are aware of all hazards related to handling and use of hydrogen. In general, hydrogen safety concerns are not more severe than those we are accustomed to with gasoline or natural gas, but they are simply different.

Strangely, most hydrogen hazards stem from the fact that hydrogen gas is odorless, colorless, and tasteless, so leaks are not detected by human senses. This is why hydrogen sensors are often used in industry to successfully detect hydrogen leaks. By comparison, natural gas is also odorless, colorless, and tasteless, but in industry mercaptans are usually added as odorants to make it detectable by people. Unfortunately, all known odorants contaminate fuel cells (a popular application for hydrogen) and are not acceptable in food applications (hydrogenation of edible oils) [1].

4.1 Physiological Hazards

Harm to people can be expressed in terms of injury or death after exposure to flames, radiant heat fluxes, extremely low temperatures, or air blast waves. Asphyxiation hazard exists if the oxygen content drops below 18 percent v/v because of hydrogen accumulation in the air. Direct skin contact with cold gaseous or liquid hydrogen leads to numbness and a whitish coloring of the skin and finally to frostbite. The risk is higher than that of liquid nitrogen because of the lower temperature and the higher thermal conductivity of hydrogen. Thus, personnel present during leaks, fires, or explosions of hydrogen mixtures with air can incur several types of injury [1-4].

4.1.1 Asphyxiation

Hydrogen is not toxic and does not pose any acute or long-term physiological hazards. The only side effect of inhalation of the gas is sleepiness and a high-pitched voice. However, asphyxiation may occur when entering a

TABLE 4.1

Stages of Asphyxiation in Presence of Hydrogen in Air

Oxygen in Air (% by Volume)	Consequences
15–19	Decreased ability to perform tasks and early symptoms in persons with heart, lung, or circulatory problems may be induced
12–15	Deeper respiration, faster pulse, poor coordination
10–12	Giddiness, poor judgment, slightly blue lips
8–10	Nausea, vomiting, unconsciousness, ashen face, fainting, mental failure
6–8	Death in 8 min. 50% death and 50% recovery with treatment in 6 min, 100% recovery with treatment in 4 to 5 min
4	Coma in 40 s, convulsions, respiration ceases, death

Source: Data from References 1–4.

region where hydrogen (as any other nontoxic gas) has displaced the air, lowering the oxygen concentration below 19.5 percent by volume. The stages of asphyxiation based on the oxygen concentration are shown in Table 4.1.

4.1.2 Thermal Burns

Thermal burns result from the radiant heat emitted by a hydrogen fire and absorbed by a person. Due to the absence of carbon and the presence of heat-absorbing water vapor created when hydrogen burns, a hydrogen fire has significantly less radiant heat compared to a hydrocarbon fire. Absorbed radiant heat is directly proportional to many factors, including exposure time, burning rate, heat of combustion, size of the burning surface, and atmospheric conditions (mainly wind and humidity). Hydrogen flames are nearly invisible in daylight and this has been the reason for fatal thermal burns suffered by victims approaching an invisible jet fire. Hydrogen fires last only one-fifth to one-tenth of the time of hydrocarbon fires. Thus, the fire damage is less severe because of:

- High burning rate resulting from rapid mixing and high propagation velocity
- High buoyant velocity
- High rate of vapor generation of liquid hydrogen

Although the maximum flame temperature of hydrogen is not much different from that of other fuels, the thermal energy radiated from the flame is only a part, for instance, of that of a natural gas flame.

The amount of damage that a burn can cause depends on its location, its depth, and how much body surface area it involves. Burns are classified on the basis of their depth in the victim's body:

- A *first-degree burn* is superficial and causes local inflammation of the skin, which is characterized by pain, redness, and a mild amount of swelling.
- A *second-degree burn* is deeper and in addition to the pain, redness, and inflammation, there is also blistering of the skin.
- A *third-degree burn* is deeper still, involving all layers of the skin, in effect killing that area of skin. Because the nerves and blood vessels are damaged, third-degree burns appear white and leathery and tend to be relatively painless.

Burns are not static and may mature. Over a few hours a first-degree burn may involve deeper structures and become second degree. Think of a sunburn that blisters the next day. Similarly, second-degree burns may evolve into third-degree burns.

In general, thermal radiation flux exposure levels may have the consequences shown in Table 4.2 [1-5].

It is interesting that the radiant heat effects on humans are a function of both the heat flux intensity and duration of exposure. Thus, it is generally accepted that harm from radiant heat has to be expressed in terms of a thermal dose unit, as given by the equation:

$$\text{Thermal Dose Unit} = I^{4/3}t \tag{4.1}$$

where I is the radiant heat flux in kW/m^2 and t is the exposure duration in seconds.

Thermal doses from ultraviolet or infrared radiation that result in first-, second-, and third-degree burns are shown in Table 4.3 [5]. As can be concluded from this table, which was based on experiments using animal skin or nuclear blast data, infrared radiation is more hazardous than that in the ultraviolet spectrum.

Based on these thermal dose levels, *dangerous dose* levels have been defined as the dose resulting in 1 percent of deaths in the exposed population.

TABLE 4.2

Radiant Heat Flux Harm Criteria for People

Thermal Radiation Intensity (kW/m^2)	Consequences on Humans
1.6	No harm for long exposures
4–5	Pain felt in 20 s; first-degree burns in 30 s
9.5	Immediate skin reactions; second-degree burns after 20 s
12.5–15	First-degree burns after 10 s; 1% lethality in 1 min
25	Significant injury in 10 s; 100% lethality in 1 min
35–37.5	1% lethality in 10 s

Source: Data from References 1–5.

TABLE 4.3

Radiation Burn Data Caused by Ultraviolet or Infrared Radiation

	Threshold Dose $(kW/m^2)^{4/3}$	
Burn Severity	**Ultraviolet**	**Infrared**
First degree	260–440	80–130
Second degree	670–1,100	240–730
Third degree	1,220–3,100	870–2,640

Source: LaChance, J., Tchouvelev, A., and Engebo, A., *International Journal of Hydrogen Energy*, 36, 2381, 2011. With permission from Elsevier.

Furthermore, *LD50* values (lethal dose to 50 percent of the exposed population) have been determined only for infrared radiation, but with a wide range of values. The Health and Safety Executive (HSE) of the United Kingdom proposes the value of 2000 $(kW/m^2)^{4/3}$ to be used in offshore and gas facilities [5].

Nevertheless, such point values are useful to be applied in quantitative risk assessment (QRA) consequence studies. Thus, the *probit function* approach appears more suitable in risk assessment of the probability of injury or fatality from a certain dose level. In probability and statistics, the probit (probability unit) function is the inverse cumulative distribution function, or quantile function, associated with the standard normal distribution. Using this function, the probability of a certain degree of harm to humans (e.g., fatality) or structures (e.g., destruction) can be obtained from the equation:

$$P\,(\text{fatality}) = 50\left[\frac{1+(Y-5)}{|Y-5|} + erf\,\frac{|Y-5|}{\sqrt{2}}\right] \tag{4.2}$$

where: Y = the value of the probit function calculated from Table 4.4, and $erf(x)$ = the error function.

The error function is a special function of sigmoid shape which occurs in probability, statistics, and partial differential equations, and is defined as:

$$\text{erf}(x) = \frac{2}{\sqrt{\pi}}\int_0^x e^{-t^2}\,dt \tag{4.3}$$

The probit functions available to be used to determine the probability of first-degree or second-degree burns or a fatality as a function of radiation heat flux are shown in Table 4.4 [5-10].

Of the equations given in Table 4.4, the Tsao and Perry probit is expected to give conservative results for exposure to hydrogen fires. The Eisenberg

TABLE 4.4

Thermal Dose Probit Functions for Humans

Probit	Probit Equation	Comment
First-degree burn (TNO) [6]	$Y = -39.83 + 3.0186 \ln [V]$[a]	Based on Eisenberg model but accounts for infrared radiation
Second-degree burn (TNO) [6]	$Y = -43.14 + 3.0186 \ln [V]$[a]	Based on Eisenberg model but accounts for infrared radiation
Fatality (Eisenberg) [7]	$Y = -38.48 + 2.56 \ln [V]$[a]	Based on nuclear data from Hiroshima and Nagasaki (ultraviolet radiation)
Fatality (Tsao and Perry) [8]	$Y = -36.38 + 2.56 \ln [V]$[a]	Eisenberg model modified to account for infrared (2.23 factor)
Fatality (TNO) [9]	$Y = -37.23 + 2.56 \ln [V]$[a]	Tsao and Perry model modified to account for clothing (14%)
Fatality (Lees) [10]	$Y = -29.02 + 1.99 \ln [V]$[b]	Accounts for clothing, based on porcine skin experiments using ultraviolet source to determine skin damage, uses burn mortality information

[a] $V = I^{4/3}t$ = thermal dose in $[(W/m^2)^{4/3}s]$.

[b] $V = F \times I^{4/3}t$ = thermal dose in $[(W/m^2)^{4/3}s]$, where $F = 0.5$ for normally clothed population and 1.0 when clothing ignition occurs.

Source: LaChance, J., Tchouvelev, A., and Engebo, A., International Journal of Hydrogen Energy, 36, 2381, 2011. With permission from Elsevier.

probit may result in lower estimations of fatalities for hydrogen fires, because it does not include the infrared spectrum. In addition, the Eisenberg probit is more suitable for hydrocarbon fires than the Tsao and Perry probit. As regards hydrogen, the best values for hydrogen fires lie between the results given by these two probit functions, but closer to the Eisenberg probit.

4.1.3 Cryogenic Burns

Cryogenic burns (frostbite) may result from contact with extremely cold fluids or cold vessel surfaces. However, to keep hydrogen, as well as other cryogenic liquids, ultra-cold today, liquid hydrogen containers are double-walled, vacuum-jacketed, super-insulated containers that are designed to vent hydrogen safely in gaseous form if a breach of either the outer or inner wall is detected. The robust construction and redundant safety features dramatically reduce the likelihood of human contact.

Frostbite is damage to tissues from freezing, due to the formation of ice crystals inside cells, rupturing and destroying them in this way. Analogous to thermal burns, frostbites are classified on the basis of their depth:

- A *first-degree injury* occurs when only the surface skin is frozen, and then the injury is called *frostnip*, starting with itching and pain. Subsequently, the skin turns white and the area becomes numb.

Frostnip generally does not lead to permanent damage. However, frostnip can lead to long-term sensitivity to heat and cold.

- A *second-degree injury* will occur when freezing continues. In this case, the skin may become frozen and hard, while the deep tissues are spared and remain soft and normal. This type of injury generally blisters 1–2 days after freezing. The blisters may become hard and blackened. However, they usually look worse than they actually are. Most of these injuries heal over 3–4 weeks, although the area may remain permanently sensitive to heat and cold.
- *Third- and fourth-degree injuries* occur when further freezing continues, leading to deep frostbite. The extremity is hard, feels woody, and use is lost temporarily, and in severe cases, permanently. The involved area appears deep purple or red, with blisters that are usually filled with blood. This type of severe frostbite may result in the loss of fingers and toes.

4.1.4 Hypothermia

Exposure to large liquefied hydrogen spills may result in hypothermia if proper precautions are not taken. Hypothermia occurs when body temperature drops below 35°C and becomes life threatening below 32.2°C. The major initial sign of hypothermia is a decrease in mental function that leads to impaired ability to make decisions, which may lead to further safety implications. Tiredness or lethargy, changes in speech, and disorientation are also typical. The affected persons will act as if they are drunk. The body gradually loses protective reflexes such as shivering, which is an important heat-generating defense. Other muscle functions also disappear so that the person cannot walk or stand. Eventually the person loses consciousness.

Recognizing hypothermia may be difficult for inexperienced persons, because the symptoms at first resemble other causes of change in mental and motor functions, such as diabetes, stroke, and alcohol or drug use. The most important thing is to be aware of the possibility and be prepared to intervene. Treatment involves slow heating of the body using blankets or other ways of increasing body warmth. Body temperature should increase by no more than a couple of degrees per hour.

4.1.5 Overpressure Injury

The effects of blast waves on humans may be direct or indirect. The principal direct effect is the sudden increase of pressure that can cause damage to pressure-sensitive organs such as the lungs and ears. Indirect effects are caused by the impact on the human body from fragments, shrapnel, and debris generated by an explosion event, by collapsing structures, or by a violent shift of the body due to the impulse generated by an explosion and

TABLE 4.5

Damage to Humans from Overpressure Events

Overpressure (kPa)	Description of Damage
Direct effects on humans	
13.8	Threshold for eardrum rupture
34.5–48.3	50% probability of eardrum rupture
68.9–103.4	90% probability of eardrum rupture
82.7–103.4	Threshold for lung hemorrhage
137.9–172.4	50% probability of fatality from lung hemorrhage
206.8–241.3	90% probability of fatality from lung hemorrhage
48.3	Threshold of internal injuries by blast
482.6–1,379	Immediate blast fatalities
Indirect effects on humans	
10.3–20.0	People knocked down by pressure waves
13.8	Possible fatality by being projected against obstacles
55.2–110.3	People standing up will be thrown a distance
6.9–13.8	Threshold of skin lacerations by missiles
27.6–34.5	50% probability of fatality from missile wounds
48.3–68.9	100% probability of fatality from missile wounds

Source: Data from References 5 and 11.

subsequent collision against a hard surface. Examples of the level of overpressure required to cause damage to humans are given in Table 4.5 [5, 11].

Nevertheless, as previously discussed for the effects of thermal radiation, blast waves from explosions will cause overpressure injury or death as a result of a combination of overpressure and duration of the effect. Again, when making assessments based on a QRA approach, one can find in the literature probit equations for many cases, including peak overpressure or impulse generated by an explosion as variables, and death from lung hemorrhage, head or whole body impact, or fragments of various sizes [5].

4.2 Physical Hazards

Among the most important physical hazards of hydrogen are those related either to its very small molecular size or to its storage generally at low temperatures.

4.2.1 Hydrogen Embrittlement

Hydrogen embrittlement (also discussed in Chapter 11, Section 11.1) is the process by which a metal becomes brittle and develops small cracks due to exposure to hydrogen. It is a long-term effect and occurs from prolonged use of a hydrogen system. Then, the mechanical properties of the metallic and nonmetallic materials of containment systems may degrade and fail resulting in spills and leaks, which will create hazards to the surroundings. Most of the damage caused in the latter case is usually incurred by ignition of the hydrogen following the rupture. Thus, all repairs and modifications to piping and equipment that handle hydrogen must be carefully engineered and tested.

The mechanisms of hydrogen embrittlement are not well understood, but certain factors are known to affect the rate of embrittlement, including hydrogen concentration, pressure, and temperature of the environment, the purity, concentration and exposure time of the hydrogen, and the stress state, physical and mechanical properties, microstructure, surface conditions, and nature of the crack front of the material (Figure 4.1) [12, 13].

4.2.1.1 Types of Embrittlement

Hydrogen embrittlement types include:

- *Environmental hydrogen embrittlement* of metals and alloys that may be plastically deformed in a gaseous hydrogen environment, resulting in increased surface cracks, loss of ductility, and decrease of fracture stress, with cracks starting at the surface.

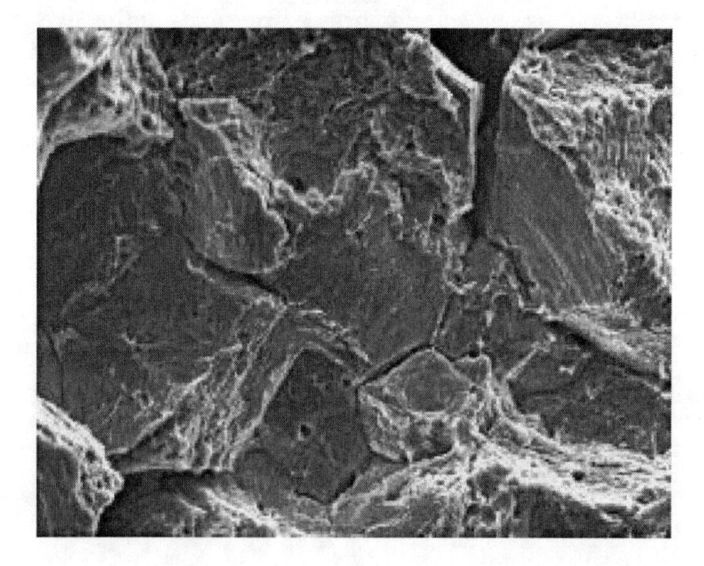

FIGURE 4.1
Scanning electron microscope image of fracture resulting from hydrogen embrittlement. This failure mode results from the absorption of hydrogen atoms into a component, which then accumulate at the grain boundaries, exerting an internal tensile stress. Magnification 2000×. (With permission from MAI Metallurgical Associates, Inc. http://metassoc.com/site/services/scanning-electron-microscopy/sem-eds-application-examples/)

- *Internal hydrogen embrittlement* caused by absorbed hydrogen, resulting in premature failure of certain metals, with cracks starting internally.
- *Hydrogen reaction embrittlement* caused by the chemical reaction of the absorbed hydrogen with one or more of the constituents of the metal to form a brittle metal hydride or methane with the carbon in steels. This phenomenon is favored by elevated temperatures.

4.2.1.2 Mechanical Properties Deterioration

Considerable deterioration of mechanical properties of many metals and alloys in the presence of hydrogen has been observed. Studies [14, 15] have demonstrated that:

- The susceptibility of a metal (or alloy) due to hydrogen increases with the strength of the metal.
- Internal and environmental hydrogen embrittlement rates are maximum in the temperature range of 200 to 300 K (-73 to 27°C), whereas hydrogen reaction embrittlement occurs at temperatures above room temperature.
- The susceptibility of steel to hydrogen embrittlement increases with hydrogen purity.
- The susceptibility to embrittlement generally increases with the tensile stress.
- Embrittlement usually results in fatigue of the metal due to crack growth.

4.2.1.3 Main Factors of Hydrogen Embrittlement

Surface and surface films: The hydrogen compatibility of metastable austenitic stainless steels such as SS type 304 depends considerably on metal surface finish. The extent of surface cracking and ductility loss may be minimized by removing the layer produced by machining. It has been found that the oxides formed naturally on the metal surfaces restrict hydrogen absorption and control the degree of embrittlement due to their lower permeability than the base metals. Synthetic surface films used to reduce hydrogen absorption should be ductile at their operating temperature. Copper and gold, which remain ductile in a broad temperature range, are usually recommended [13].

Effect of electrical discharge machining: Hydrogen embrittlement can be increased by electrical discharge machining commonly used for producing holes, notches, slots, or other cavities in metals. In this way, electrical discharges can introduce hydrogen into a machined component. Hydrogen is produced by the decomposition of the dielectric fluid (usually oil or kerosene) used when it is ionized by the electrical discharge.

Effect of trapping sites: Hydrogen may be trapped at sites within the structure of the metal, including dislocations, grain and phase boundaries, interstitial or vacancy clusters, voids or gas bubbles, oxygen or oxide inclusions, carbide particles, and other material defects. Trapping is most marked at low temperatures and is important at ambient temperatures where hydrogen embrittlement is also distinct.

4.2.1.4 Hydrogen Embrittlement Control

In general, stainless steel is more resistant to hydrogen embrittlement than ordinary steels, and both pure aluminum and many aluminum alloys are even more resistant than stainless steel if the gas is dry. As also described at various points in Chapters 7, 10, and 11, all components of hydrogen fuel systems must be constructed of materials known to be compatible with hydrogen [12].

Oxide coatings, elimination of stress concentrations, additives to hydrogen, proper grain size, and careful alloy selection are among the measures to prevent embrittlement. In addition, the following measures [13] can be used to effectively eliminate hydrogen embrittlement in metals:

- Use of aluminum as structural material due to its low susceptibility to hydrogen.
- Design of components made of medium-strength steel (for gaseous hydrogen) and stainless steel (for liquid hydrogen) on the basis of increased thickness, surface finish, and proper welding techniques.
- In the absence of data, design of metal components assuming a substantial (up to fivefold) decrease in resistance to fatigue.
- Not using cast iron and hydride-forming metals and alloys as structural materials for hydrogen service.
- Taking into consideration that, in general, exposure temperatures below room temperature retard hydrogen reaction embrittlement.
- Taking into consideration that, in general, environmental and internal hydrogen embrittlement are increased in the temperature range of 200 to 300 K (-73 to 27°C).

4.2.2 Thermal Stability of Structural Materials

4.2.2.1 Low-Temperature Mechanical Properties

For safety reasons, selection of structural materials is based primarily on their mechanical properties, such as yield and tensile strength, ductility, and impact strength. The materials should address minimum design values of these properties over the entire temperature range of operation taking into account nonoperational conditions such as a hydrogen fire. The material properties should be stable without phase changes in the crystalline structure with time or repeated thermal cycling [13].

The main considerations of metal and alloy behavior at low temperature conditions are the following:

- The transition from ductile to brittle behavior at low temperatures.
- Certain unconventional modes of plastic deformation encountered at very low temperatures.
- Mechanical and elastic property changes due to phase transformations in the crystalline structure at low temperatures.

Especially for the selection of a material for liquefied hydrogen service, the main thermal properties to be considered are low-temperature embrittlement and thermal contraction.

4.2.2.2 Low-Temperature Embrittlement

At lower temperatures and especially at cryogenic temperatures, many materials change from ductile to brittle. This change may lead to failure of a hydrogen storage vessel or pipe and cause an accident. Such an accident occurred in Cleveland in 1944 due to low-temperature embrittlement of a liquid natural gas (LNG) storage vessel made of 3.5 percent nickel steel with a capacity of 4,248 m^3. The vessel ruptured and released 4,163 m^3 of LNG, which ignited after having spread into nearby storm sewers. As a domino effect, a nearby storage vessel collapsed from the fire and spilled its contents, which also burned, with flames up to about 850 m height. The accident resulted in 128 deaths, 200 to 400 injuries, and an estimated U.S. $6,800,000 (1944 dollars) in property damage [13].

The ductility of a material is commonly determined by the Charpy impact test. The results of the Charpy impact test at various temperatures for several materials are shown in Figure 4.2 [13].

The gradual loss of ductility of 9 percent nickel steel is clearly shown in this figure, as well as the stepwise embrittlement of 201 stainless steel at temperatures below 280 K (7°C) and C1020 carbon steel below 120 K (-153°C). This indicates that these materials should not be used for liquid hydrogen storage. The ductility of aluminum 2024-T4 does not change considerably at low temperatures, but its low strength should be taken into account in storage vessel design. Strangely, the Charpy impact strength of 304 stainless steel increases when the temperature decreases, indicating that this is a suitable material for the construction of liquefied hydrogen storage vessels and pipes.

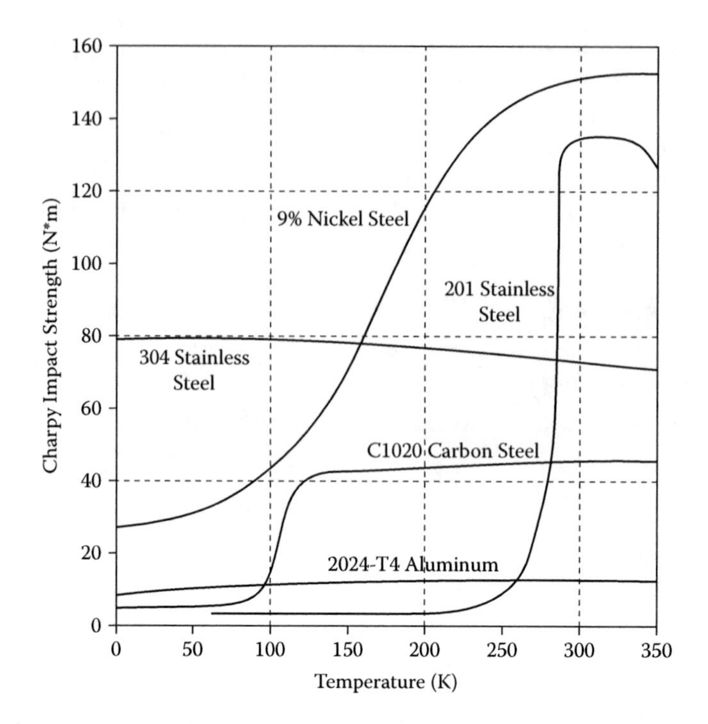

FIGURE 4.2
Charpy impact strength as a function of temperature for various materials. (From *Guide to Safety of Hydrogen and Hydrogen Systems,* American Institute of Aeronautics and Astronautics, American National Standard ANSI/AIAA G-095-2004. With permission from the American Institute of Aeronautics and Astronautics.)

The difference between the yield and tensile strengths of a material may also be used to measure its ductility. As a rule, a material is considered ductile when these two values differ considerably. Thus, in the case of 5986 aluminum, the values of yield and tensile strengths stand out as the temperature decreases in the cryogenic region, indicating that this is a suitable material for liquefied hydrogen. However, these values are similar in the cryogenic region for AISI 430 stainless steel, indicating its unsuitability for use with liquefied hydrogen.

4.2.2.3 Thermal Contraction

Different structural materials have different thermal contraction coefficients, meaning that accommodations should be made for their different dimensions at cryogenic temperatures. If not, problems associated with safety (e.g., leaks) may arise. In general, the contraction of most metals from room temperature (300 K) to a temperature close to the liquefaction temperature of hydrogen (20 K) is less than 1 percent, whereas the contraction for most common structural plastics is from 1 to 2.5 percent [13].

4.3 Chemical Hazards

4.3.1 General Considerations and Accident Statistics

As shown in Table 3.1, major emphasis should be placed on containment, detection, and ventilation because the minimum energy of gaseous hydrogen (GH_2) for ignition in air at atmospheric pressure is about 0.02 mJ. Experience has shown that escaped hydrogen is very easily ignited.

For hydrogen (as fuel) to catch fire, it needs to be mixed with air (oxygen as oxidant) and the mixture must be within the flammability limits. In addition, and in accordance with the familiar fire triangle, an ignition source with sufficient energy must be present for hydrogen to burn. Leaks occur for every gas and especially for hydrogen, which even the best efforts cannot contain.

Safety measures include elimination of all likely sources of ignition. An investigation of industrial accidents has shown that 53 percent of them occurred because of leaks, off-gassing, and equipment rupture, as listed in Table 4.6 [17]. Purging or vent-exhaust incidents account for 15 percent and the remaining 32 percent were other types of incidents. In ammonia plants, most of the accidents are due to leaking of gaskets and valve packing, as shown in Table 4.7 [18]. The aerospace industry has paid a great toll, with 107 hydrogen incidents in 1974 alone, of which 87 involved release of GH_2 or LH_2, as shown in Table 4.8 [19]. When accidents were not caused by equipment failure, they primarily occurred when procedures were not prescribed or when prescribed procedures were not followed. Ignition sources responsible for these incidents were electrical short circuits and sparks (25 percent), static charges (18 percent), welding or cutting torches, metal fracture, gas impingement and the rupture of safety disks (3 to 6 percent), as accident reports have shown [2].

TABLE 4.6

Industrial Hydrogen Accidents

Category	Number of Accidents	Percentage Total Accidents
Undetected leaks	32	22
Hydrogen-oxygen off-gassing explosions	25	17
Piping and pressure vessel ruptures	21	14
Inadequate inert gas purging	12	8
Vent and exhaust system incidents	10	7
Hydrogen-chlorine incidents	10	7
Others	35	25
Total	145	100

Source: Adapted from Zalosh, R.G., and Short, T.P., *Compilation and Analysis of Hydrogen Accident Reports*, C00-4442-4, Department of Labor, Occupational Safety and Health Administration, Factory Mutual Research Corp., Norwood, 1978.

TABLE 4.7

Hydrogen Accidents in Ammonia Plants

Classification	Number	Percentage Total Accidents
Gaskets:	46	37
Equipment flanges	23	18
Piping flanges	16	13
Valve flanges	7	6
Valve packing	10	8
Oil leaks	24	19
Transfer header	9	7
Auxiliary boiler	8	6
Primary reformer	7	6
Cooling tower	3	2
Electrical	2	2
Miscellaneous	16	13
Total	125	100

Source: Williams, G.P., *Chemical Engineering Progress*, 74, 9, 1978. With permission.

TABLE 4.8

Hydrogen Accidents in the Aerospace Industry

Description	Accidents Involving Release of Hydrogen	Percentage of Total Accidents
Accidents involving release of liquid or gaseous hydrogen	87	81
Location of hydrogen release:		
To atmosphere	71[a]	66
To enclosures (piping, containers, etc.)	26[a]	24
Ignition of hydrogen releases:		
To atmosphere	44	41
To enclosures	24	22

[a] Hydrogen was released to both locations in 10 accidents.

Source: Ordin, P.M., A review of hydrogen accidents and incidents in NASA operations, in *9th Intersociety Energy Conversion Engineering Conference*, American Society of Mechanical Engineers, New York, 1974. With permission.

4.3.2 Flammability of Hydrogen

The flammability limits of mixtures of hydrogen with air, oxygen, or other oxidizers depend on the ignition energy, temperature, pressure, presence of diluents, and size and configuration of the equipment, facility, or apparatus. Such a mixture may be diluted with either of its constituents until its concentration shifts below the lower flammability limit (LFL) or above the upper flammability limit (UFL). The flammability limits of hydrogen mixtures with either air or oxygen widen for upward flame propagation and narrow for downward flame propagation.

Mixtures of LH_2 and liquid oxygen (LOX) or solid oxygen as oxidizer are not hypergolic. In accidental fires of these mixtures during the mixing process the system caught fire because the required ignition energy is very small [2]. Nevertheless, LH_2 and liquid or solid oxygen can detonate when initiated by a shock wave.

The flammability limits of hydrogen in dry air for upward propagation in tubes at 101.3 kPa (1 atm) and ambient temperature range from 4.1 percent (LFL) to 74.8 percent (UFL). The flammability limits of hydrogen in oxygen for upward propagation in tubes at 101.3 kPa (1 atm) and ambient temperature are from 4.1 percent (LFL) to 94 percent (UFL). With a reduction in pressure below 101.3 kPa, the range of flammability limits narrows, as shown in Table 4.9 [2].

TABLE 4.9

Flammability Limits of Hydrogen-Air and Hydrogen-Oxygen Mixtures

	Hydrogen Content (vol%)					
	Upward Propagation		**Downward Propagation**		**Horizontal Propagation**	
Conditions	**LFL**	**UFL**	**LFL**	**UFL**	**LFL**	**UFL**
Hydrogen in air and oxygen at 101.3 kPa (1 atm)						
H2 + air:						
Tubes	4.1	74.8	8.9	74.5	6.2	71.3
Spherical vessels	4.6	75.5	—	—	—	—
H2 + oxygen	4.1	94.0	4.1	92.0	—	—
Hydrogen plus inert gas mixtures at 101.3 kPa (1 atm)						
H_2 + He + 21 vol% O_2	7.7	75.7	8.7	75.7	—	—
H_2 + CO_2 + 21 vol% O_2	5.3	69.8	13.1	69.8	—	—
H_2 + N_2 + 21 vol% O_2	4.2	74.6	9.0	74.6	—	—

Hydrogen in air at reduced pressure with a 45 mJ ignition source

	25 cm Tube		**2 L Sphere**	
Pressure (kPa)	**LFL**	**UFL**	**LFL**	**UFL**
20	~4	~56	~5	~52
10	~10	~42	~11	~35
7	~15	~33	~16	~27
6	20–30		20–25 (at 6.5 kPa)	

Note: Dashes indicate no information available.

Source: *Guide to Safety of Hydrogen and Hydrogen Systems*, American Institute of Aeronautics and Astronautics, American National Standard ANSI/AIAA G-095-2004, 2004. With permission.

4.3.2.1 Hydrogen-Air Mixtures

The lowest pressure at which a low-energy ignition source ignites a hydrogen-air mixture is approximately 6.9 kPa (0.07 atm) at a hydrogen concentration between 20 and 30 percent by volume [2].

Hydrogen-air mixtures ignited by a 45 mJ spark ignition source at 311 K (38°C) environment temperature have an LFL equal to 4.5 percent by volume over the pressure range 34.5 to 101.3 kPa (0.34 to 1 atm). An increasingly higher LFL was required to obtain combustion below 34.5 kPa (0.34 atm). The lowest pressure at which a low-energy ignition source could inflame the mixture was 6.2 kPa (0.06 atm) at a hydrogen-air mixture of between 20 and 30 percent by volume of hydrogen [20]. Yet, using a strong ignition source, the lowest pressure at which ignition can occur is 0.117 kPa (0.0012 atm).

For downward propagation, the LFL of hydrogen-air decreases from 9.0 to 6.3 percent by volume hydrogen and the UFL increases from 75 to 81.5 percent by volume of hydrogen when the temperature is increased from 290 to 673 K (17 to 400°C) at a pressure of 101.3 kPa (1 atm).

Compared to methane-air mixtures, hydrogen-air mixtures are more hazardous with respect to accidental ignition due to the much wider flammability limits and the order of magnitude lower ignition energy, as shown in Figure 4.3 [21].

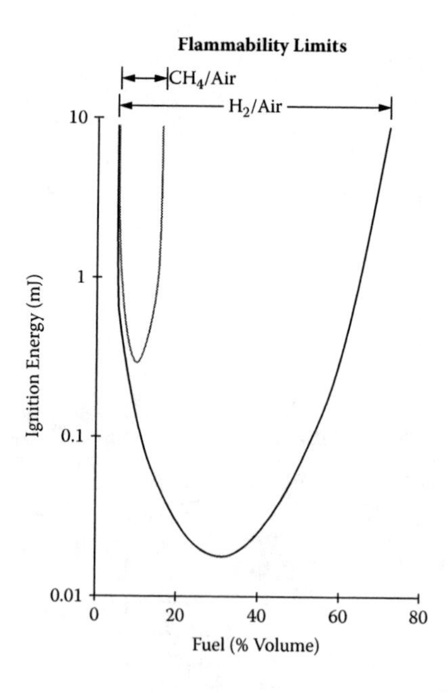

FIGURE 4.3
Minimum ignition energy of hydrogen-air and methane-air mixtures at a pressure of 101.3 kPa (1 atm) and a temperature of 298 K (25°C). (From Fisher, M., *International Journal of Hydrogen Energy*, 11(9), 593–601, 1986. With permission from Elsevier.)

4.3.2.2 Hydrogen-Oxygen Mixtures

The flammability limits for hydrogen-oxygen mixtures at 101.3 kPa (1 atm) range from 4 to 94 percent by volume of hydrogen for upward propagation in tubes. Reduced pressures increase the LFL [2]. The lowest pressure observed for ignition is 57 kPa (0.56 atm) at a hydrogen concentration of 50 percent by volume when a high-energy ignition source was used.

At elevated pressures, the LFL does not change with pressure up to 12.4 MPa (122 atm), while at 1.52 MPa (15 atm) the UFL is 95.7 percent by volume of hydrogen [2].

The LFL decreases from 9.6 to 9.1 percent by volume of hydrogen and the UFL increases from 90 to 94 percent by volume of hydrogen, when the temperature rises from 288 to 573 K (15 to 300°C).

4.3.2.3 Effects of Diluents

The flammability limits for hydrogen-oxygen-nitrogen mixtures are shown in Figure 4.4 [2]. Table 4.9 shows flammability limits for GH_2 and gaseous oxygen (GOX) with equal concentrations of added inert gases (helium, carbon dioxide, and nitrogen). Table 4.10 shows the qualitative effect of helium, carbon dioxide, nitrogen, and argon diluents for various tube sizes [2]. Argon is the least effective diluent in reducing the flammable range for hydrogen in air.

The effects of helium, carbon dioxide, nitrogen, and water vapor on the flammability limits of hydrogen in air are shown in Figure 4.5 [2].

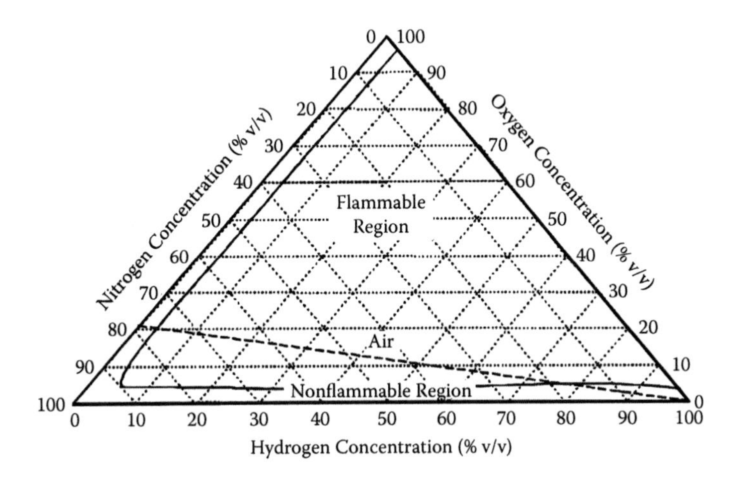

FIGURE 4.4
Flammability limits of hydrogen-oxygen-nitrogen mixtures at a pressure of 101.3 kPa (1 atm) and a temperature of 298 K (25°C). (From *Guide to Safety of Hydrogen and Hydrogen Systems*, American Institute of Aeronautics and Astronautics, American National Standard ANSI/AIAA G-095-2004. With permission from the American Institute of Aeronautics and Astronautics.)

TABLE 4.10

Effects of Equal Concentrations of Diluents on Flammable Range for Hydrogen in Air

Tube Diameter (cm)	Rating of Diluents at Reducing Flammable Range
Wide tubes	$CO_2 < N_2 < He < Ar$
2.2	$CO_2 < He < N_2 < Ar$
1.6	$He < CO_2 < N_2 < Ar$

Source: *Guide to Safety of Hydrogen and Hydrogen Systems,* American Institute of Aeronautics and Astronautics, American National Standard ANSI/AIAA G-095-2004, 2004. With permission.

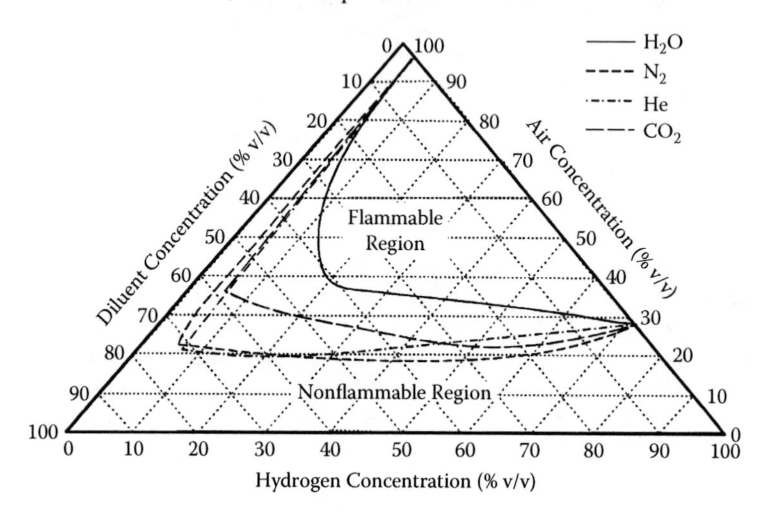

FIGURE 4.5
Effects of N_2, He, CO_2, and H_2O diluents on flammability limits of hydrogen in air at 101.3 kPa (1 atm). (The effects of N_2, He, and CO_2 are at 298 K [25°C] and H_2O is at 422 K [149°C].) (From *Guide to Safety of Hydrogen and Hydrogen Systems,* American Institute of Aeronautics and Astronautics, American National Standard ANSI/AIAA G-095-2004, 2004. With permission from the American Institute of Aeronautics and Astronautics.)

Measurements were performed at 298 K (25°C) and 101.3 kPa (1 atm) except for the water vapor measurements, which were performed at 422 K (149°C). Water was the most effective diluent in reducing the flammability range of hydrogen in air.

4.3.2.4 Effects of Halocarbon Inhibitors

The effects of halocarbon inhibitors on the flammability limits of hydrogen-oxygen mixtures are shown in Figure 4.6 [22].

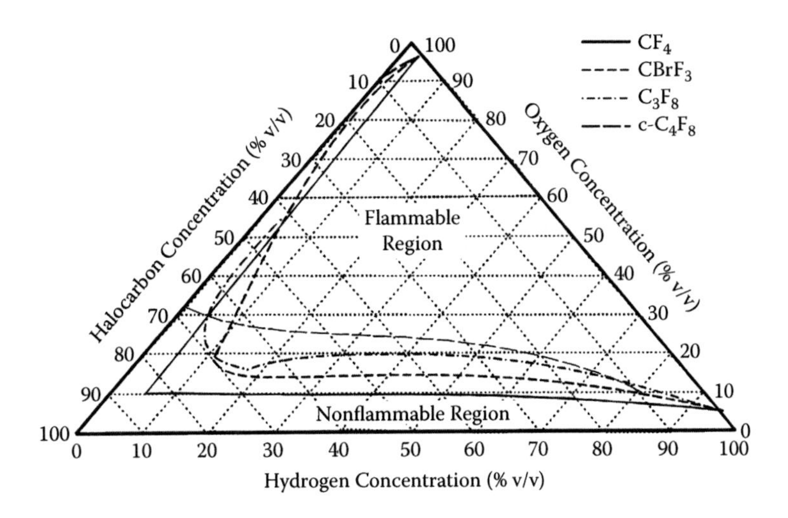

FIGURE 4.6
Effects of halocarbon inhibitors on flammability limits of hydrogen-oxygen mixtures at a pressure of 101.3 kPa (1 atm) and a temperature of 298 K (25°C). (From *Guide to Safety of Hydrogen and Hydrogen Systems*, American Institute of Aeronautics and Astronautics, American National Standard ANSI/AIAA G-095-2004, 2004. With permission from the American Institute of Aeronautics and Astronautics.)

The effect of N_2, CH_3Br, and $CBrF_3$ on the extinguishment of hydrogen diffusion flames in air is compared in Table 4.11 [23]. The halocarbon inhibitors were more effective when added to the air stream; nitrogen was more effective when added to the fuel stream.

4.3.2.5 Autoignition Temperature

The autoignition temperature of hydrogen in air differs only slightly from that in oxygen. It is interesting to note that autoignition temperatures in both cases depend not only on GH_2 concentration and pressure but the surface treatment of containers as well. So the autoignition temperatures found in the literature are very dependent on the system and values should be applied only to similar systems. At 101.3 kPa (1 atm) the range of reported autoignition temperatures for stoichiometric hydrogen in air is from 773 to 850 K (500 to 577°C), while in stoichiometric concentration with oxygen it is from 773 to 833 K (500 to 560°C). At pressures from 20 to 50 kPa (0.20 to 0.49 atm), GH_2-air ignitions have occurred at 620 K (347°C).

4.3.2.6 Quenching Gap in Air

Quenching distance is the passage gap dimension required to prevent the propagation of an open flame through a flammable fuel-air mixture that fills

TABLE 4.11

Inhibitor for Extinction of Hydrogen Diffusion Flames

Inhibitor	Concentration at Flame Extinction (vol%)
Added to air:	
Nitrogen	94.1
CH_3Br	11.7
$CBrF_3$	17.7
Added to fuel:	
Nitrogen	52.4
CH_3Br	58.1
$CBrF_3$	56.6

Source: Creitz, E.C., *Journal of Research for Applied Physics and Chemistry*, 65, 389, 1961. With permission.

the passage. In parallel-plate configuration, the quenching gap is defined as the spark gap between two flat electrodes at which ignition of combustible fuel-air mixtures is suppressed.

The quenching gap for hydrogen is 0.6 mm in air at normal temperature and pressure (NTP). Nevertheless, this value depends on the temperature, pressure, and composition of the combustible gas mixture, as well as the electrode configuration.

Faster burning gases generally have smaller quenching gaps. Consequently, flame arresters for faster burning gases require smaller apertures. The lowest quenching distance found in the literature for hydrogen is 0.076 mm [24]. There are three major considerations in determining the quenching distance for a gaseous fuel such as hydrogen: ignition energy, mixture composition, and pressure.

Quenching gap depends on the *ignition energy*. For instance, a low ignition energy of the order of 0.001 mJ corresponds to a small gap of the order 0.01 cm. On the contrary, a high ignition energy of 10 mJ corresponds to a larger gap of 1 cm [2].

Pressure and *composition* also affect the quenching distance with this parameter increasing rapidly at very low pressures. The effect of *mixture ratio* is less well known. Nevertheless, quenching distance appears to be constant for a given pressure between the UFL and LFL. Since specific values for hydrogen-air mixtures are not available, the effect of pressure as a function of tube diameter for deflagration and detonation of acetylene-air mixtures can be used as a guideline [2].

4.3.3 Ignition Sources

All ignition sources such as open flames, electrical equipment, or heating equipment should be eliminated or safely isolated in buildings or special

TABLE 4.12

Potential Ignition Sources

Electrical Sources	Mechanical Sources	Thermal Sources	Chemical Sources
Static discharge	Mechanical impact	Open flame	Catalysts
Static electricity (e.g. two-phase flow)	Tensile rupture	Hot surface	Reactants
Static electricity (e.g. flow with solid particles included)	Friction and galling	Personnel smoking	
Electric arc	Mechanical vibration	Welding	
Lightning	Metal fracture	Exhaust from combustion engine	
Charge accumulation		Resonance ignition	
Electric charge generated by equipment operation		Explosive charge	
Electrical short circuits		High-velocity jet heating	
Electrical sparks		Shock wave from tank rupture	
Clothing (static electricity)		Fragment from bursting tank	

Note: This list should not be viewed as a complete list. Be alert for other possible ignition sources.

Source: *Guide to Safety of Hydrogen and Hydrogen Systems*, American Institute of Aeronautics and Astronautics, American National Standard ANSI/AIAA G-095-2004, 2004. With permission.

rooms containing hydrogen systems. Operations should be conducted as if unforeseen ignition sources could occur.

Potential ignition sources for hydrogen systems are shown in Table 4.12 [2]. The ignition of GH_2-air mixtures usually results in deflagration, with explosion consequences significantly less severe than if detonation results. It is possible that in a confined or partially confined enclosure a deflagration can evolve into a detonation, a phenomenon known as deflagration-to-detonation-transition (DDT). The geometry and flow conditions (turbulence) have a strong effect on DDT.

4.3.3.1 Sparks

Electrical sparks can be the result of electrical discharges between objects having different electrical potentials, such as breaking electrical circuits or discharges of static electricity.

Static electricity sparks can ignite hydrogen-air or hydrogen-oxygen mixtures. Static electricity is caused by many common articles, such as hair or fur when combed or stroked, or an operating conveyor belt. People can generate high-voltage charges of static electricity on themselves when walking on synthetic carpet or dry ground, moving while wearing synthetic clothing, sliding on automobile seats, or combing their hair. The flow of GH_2 or LH_2 in ducts or turbulence in containers can generate charges of static electricity, as with any other nonconductive liquid or gas. In addition, static charges may be induced during electrical storms [25–27].

Hard objects coming into shearing contact with each other can cause *friction sparks,* as in the cases of metal striking metal, metal striking stone, or stone striking stone. Friction sparks are particles of burning material, initially heated by the mechanical energy of friction and impact that have been sheared off as a result of contact. Sparks from hand tools have normally low energy, while mechanical tools such as drills and pneumatic chisels can generate high energy sparks.

Hard objects striking each other can also produce *impact sparks.* Impact sparks are usually produced by impact on a quartzitic rock, such as the sand in concrete, resulting in throwing off of small particles of the impacted material.

Minimum spark energy for ignition is defined as the minimum spark energy required to ignite the most easily ignitable concentration of fuel in air and oxygen. The minimum spark energies of hydrogen in air are 0.017 mJ at 101.3 kPa (1 atm), 0.09 mJ at 5.1 kPa (0.05 atm), and 0.56 mJ at 2.03 kPa (0.02 atm). The minimum spark energy required for ignition of hydrogen in air is considerably less than that for methane (0.29 mJ) or gasoline (0.24 mJ). Nevertheless, the ignition energy for all three fuels is sufficiently low that ignition in air is rather certain in the presence of any weak ignition source, such as sparks, matches, hot surfaces, open flames, or even a weak spark caused by the discharge of static electricity from a human body.

4.3.3.2 Hot Objects and Flames

Objects at temperatures from 773 to 854 K (500 to 581°C) can ignite hydrogen-air or hydrogen-oxygen mixtures at atmospheric pressure. Substantially cooler objects, about 590 K (317°C), can also ignite these mixtures after prolonged contact at less than atmospheric pressure. Open flames can easily ignite hydrogen-air mixtures.

In cases where ignition sources are a required part of hydrogen use, provisions should be made to adequately contain any resulting deflagration or detonation. For instance, a combustor or engine should not be operated in atmospheres containing hydrogen without well-dispersed water sprays in its exhaust. Experience has shown that multiple bank sprays will partially suppress the detonation pressures and reduce the number and temperature of ignition sources in an exhaust system. Water sprays should not, however, be relied on as a means of avoiding detonations. Carbon dioxide may be used with the water spray to further reduce hazards.

4.3.4 Explosion Phenomena

4.3.4.1 Terms and Definitions

The general term *explosion* corresponds to a rapid energy release and pressure rise and can occur in both reactive and nonreactive systems. Pressurized gas vessel failures are typical examples of nonreactive explosions, whereas premixed gases of fuel and oxidizer permit a rapid energy release by chemical reactions in reactive systems. The term explosion is sometimes used interchangeably in the literature for any kind of violent pressure rise such as deflagration and detonation. Others limit the term explosion only to the bursting of an enclosure or a container due to the development of internal pressure from a deflagration [28].

Gas explosions do not necessarily require the transmission of a wave through the explosion source medium, although explosions in most cases involve some kind of wave such as a deflagration or detonation wave. For instance, in *volumetric explosions*, an explosive mixture contained in a vessel can be heated to a high enough temperature for fast reactions to occur [29].

Deflagration is the phenomenon in which the flame front moves through a flammable mixture in the form of a subsonic wave with respect to the unreacted medium. The terms deflagration, flash fire, combustion, flame, and burn are sometimes found in the literature as being used interchangeably.

Detonation occurs when the flame front coupled to a shock wave propagates through a detonable mixture in the form of a supersonic wave with respect to the unreacted medium. The thermal energy of the reaction sustains the shock wave and the shock wave compresses the unreacted material to sustain the reaction. In general, a detonation propagates two to three orders of magnitude faster than a deflagration and results in pressures at the detonation front 15 to 20 times higher than the initial pressure [29]. This explains why a detonation has much greater potential for causing personnel injury or equipment damage than a deflagration.

For gaseous mixtures exploding in the open, the term *unconfined vapor cloud explosion* (UVCE) is used, whereas when they explode in confined spaces the term *confined vapor cloud explosion* (CVCE) is used. A recent development has been the use of the term *vapor cloud explosion* or VCE for either the confined or unconfined scenario. In these cases, either a detonation or a deflagration takes place depending on the factors mentioned above. In a very lean or rich fuel mixture, the flame front travels in the cloud at a low velocity and insignificant pressure increase, a phenomenon known as a *flash fire.*

Another explosive phenomenon encountered in the storage of liquefied hydrogen under pressure is the *boiling liquid expanding vapor explosion* (BLEVE), which occurs from the sudden release of a large mass of pressurized liquid into the atmosphere. A typical primary cause is an external flame impinging on the shell of a vessel above the liquid level, thus weakening the shell and resulting in sudden rupture. Then the content of the vessel is released into the atmosphere and, if flammable, ignites forming a nearly

FIGURE 4.7
The BLEVE (boiling liquid expanding vapor explosion) is the worst possible outcome when a tank holding a pressure liquefied gas, such as propane or hydrogen, fails due to fire contact or impact. (From Google images)

spherical burning cloud, the so-called *fireball* (Figure 4.7). During burning, the combustion energy is emitted mostly in the form of radiant heat. The inner core of the cloud consists of almost pure fuel, whereas the outer layer, where the ignition first occurs, is a fuel-air mixture within the flammable limits. As buoyancy forces of the hot gases begin to dominate, the burning cloud rises and becomes more spherical in shape.

The mechanism of a BLEVE is based on the fact that in the absence of nucleation sites in a liquid, such as impurities, crystals or ions, its boiling point can be exceeded without boiling, thus resulting in superheating of the liquid. Yet, there is a limit above which the liquid cannot be further superheated at a given pressure and when this limit is reached, microscopic vapor bubbles are formed spontaneously without nucleation sites. The very high nucleation rates accompanying this phenomenon result in the formation of a shock wave transmitted through the evaporating liquid. Superheat limit state, critical properties, and nucleation rates for hydrogen and some other cryogenic materials are shown for comparison in Table 4.13 [30, 31].

TABLE 4.13

Superheat Limit State, Critical Properties, and Nucleation Rates for Hydrogen and Some Other Materials

Substance	Critical Properties Temperature (K)	Critical Properties Pressure (bar)	Superheat Limit State Temperature (K)	Superheat Limit State Pressure (bar)[a]	Normal Boiling Point (K)	Nucleation Rate (nuclei/ cm^3-s)
Hydrogen	32.98	12.93	27.8	5.54	20.3	10^{-2}
Methane	190.56	45.92	167.6	21.3	111.6	10^5
Ethane	305	49.9	269.2	21.7	184	10^5
Propane	370	43.6	326.4	18.4	231	10^5
Water	647	218	553.0	64.1	373	10^{21}

[a] Wolfram Alpha Computational Knowledge Engine was used to calculate these values.
Source: Data from References 30 and 31.

When using Van der Waals equation of state, the superheat limit temperature, T_{sl}, and critical temperature, T_c, have been correlated with the equation [30, 32]:

$$T_{sl} = 0.84\ T_c \tag{4.4}$$

or, when using the Redlich-Kwong equation of state, with the equation:

$$T_{sl} = 0.895\ T_c \tag{4.5}$$

It is obvious from Table 4.13 that the nucleation rate for hydrogen is millions of times lower than for other cryogenic gases and 10^{23} times lower than that of water. So in this respect, occurrence of a shockwave is quite improbable during a sudden loss of containment of liquefied hydrogen compared to other liquefied flammable gases. Furthermore, vessels containing liquefied hydrogen are much safer than high-pressure steam boilers, which have been involved in many catastrophic explosions due to the BLEVE phenomenon (although with no possibility of a subsequent fireball of steam).

4.3.4.2 Detonation Limits

The maximum and minimum concentrations of a gas, vapor, mist, spray, or dust in the air or other gaseous oxidant for a stable detonation to occur are the so-called upper and lower *detonation limits*. These limits depend on the size and geometry of the surroundings as well as other factors. Therefore, detonation limits found in the literature should be used with caution. Detonation limits are sometimes confused with deflagration limits and the term explosive limits is then used inconsiderately [33].

As shown in Table 3.1, detonable concentrations are narrower than flammable concentrations for hydrogen as well as for all other gases. The worst case during an accidental release is reaching a near-stoichiometric concentration of a flammable gas and encountering an ignition source, with the end result being a detonation. The occurrence of such detonable concentrations is quite infrequent in an enclosed area and rather unlikely in the open air for most gases that have lower and upper detonation limits very close to each other. Unfortunately, with some gases, including hydrogen, this concentration span is quite broad and the detonability range is easily attained leading to much more serious accidents than those arising from gases with narrow detonation ranges [34].

The higher molecular diffusion of GH_2 shown in Table 3.1 compared to other gases means that any release will quickly mix with the surrounding air, reach a hazardous range, and then come down to the safe region below the lower flammability limit. In a LH_2 leak, fast vaporization occurs first and molecular diffusion in the air follows. On the other hand, the positive buoyancy of GH_2 can move a release to either higher floors in a building endangering these areas or to safe heights in the open air.

Detonation limits for any fuel-oxidizer mixture depend, among other factors, on the nature and dimensions of confinement. The minimum dimensions for detonation of hydrogen-air mixtures at 101.3 kPa (1 atm) and 298 K (25°C) in different degrees of confinement are shown in Figure 4.8 [35].

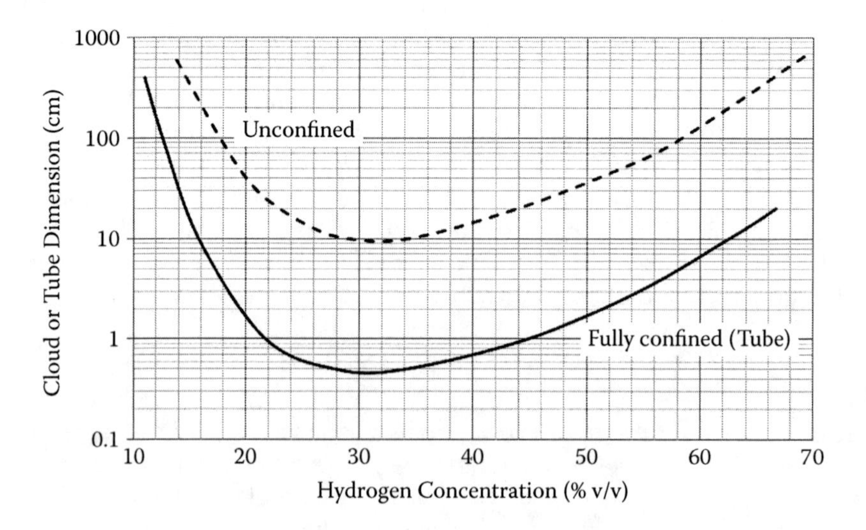

FIGURE 4.8
Minimum dimensions of GH_2-air mixtures for detonation at 101.3 kPa (1 atm) and 298 K (25°C). (From *Guide to Safety of Hydrogen and Hydrogen Systems*, American Institute of Aeronautics and Astronautics, American National Standard ANSI/AIAA G-095-2004, 2004. With permission from the American Institute of Aeronautics and Astronautics.)

4.3.4.3 Ignition Energy

The *minimum ignition energy* for a detonation to occur depends on the hydrogen concentration in the detonable mixture. Thus, the minimum ignition energy is equivalent to about 1 gram of tetryl for hydrogen-air mixtures at the stoichiometric composition (possessing the highest sensitivity and leading to direct detonation), whereas this increases to several tens of grams of tetryl for very rich or very lean mixtures (close to the detonation limits) [2].

The ignition energy required for a stable detonation is large for lean or rich mixtures (i.e., far from the stoichiometric concentration). On the other hand, it is possible to produce overdriven detonations when a large ignition energy is used.

Stable detonation parameters such as detonation temperature and pressure can be calculated by the Chapman–Jouget approach using the Gordon–McBride computer code [36].

The dependence of detonation temperature and pressure on hydrogen concentration for hydrogen-air mixtures and hydrogen-oxygen mixtures is shown in Figure 4.9 and Figure 4.10, respectively. Maximum values of approximately 3000 K and 1600 kPa for near-stoichiometric hydrogen-air mixtures are observed, whereas these maxima rise to about 3800 K and 2000 kPa for stoichiometric hydrogen-oxygen mixtures.

4.3.4.4 Detonation Cell Size

A characteristic parameter related to detonation is the *detonation cell size* (λ). The wave front in a detonation is not planar but is composed of reaction cells, as verified by the pattern left by a detonation on a smoked plate (Figure 4.11).

A two-dimensional illustration of the actual structure is given in Figure 4.12 according to the Zeldovich, von Neumann, and Döring (ZND) model.

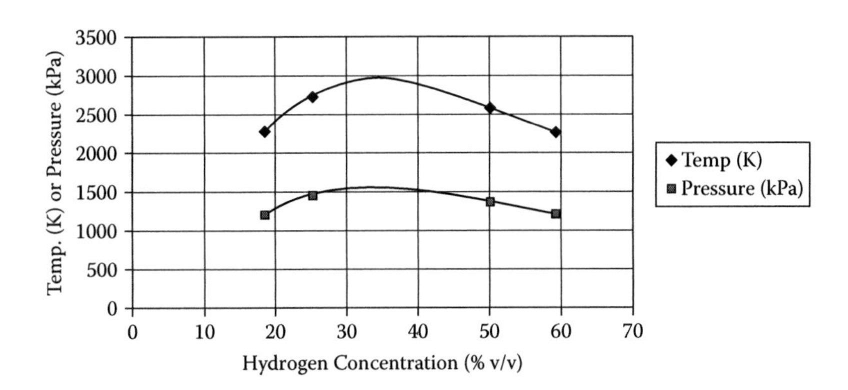

FIGURE 4.9
Detonation pressure and temperature of hydrogen-air mixtures starting from 101.3 kPa (1 atm) and 298 K (25°C). Chapman–Jouguet calculations using the Gordon–McBride code [36].

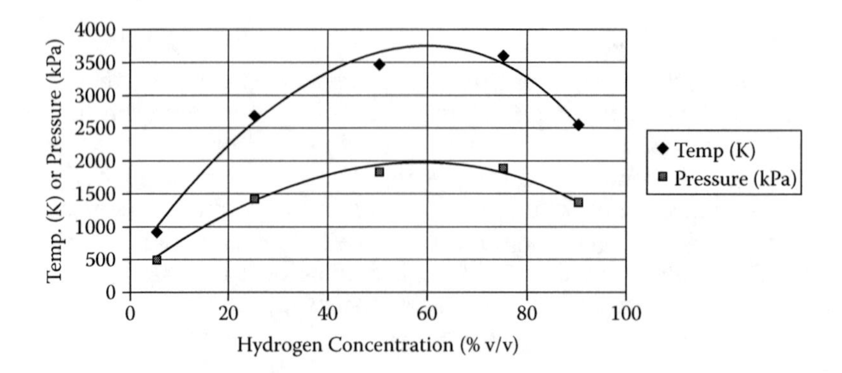

FIGURE 4.10
Detonation pressure and temperature of hydrogen-oxygen mixtures starting from 101.3 kPa (1 atm) and 298 K (25°C). Chapman–Jouguet calculations using the Gordon–McBride code [36].

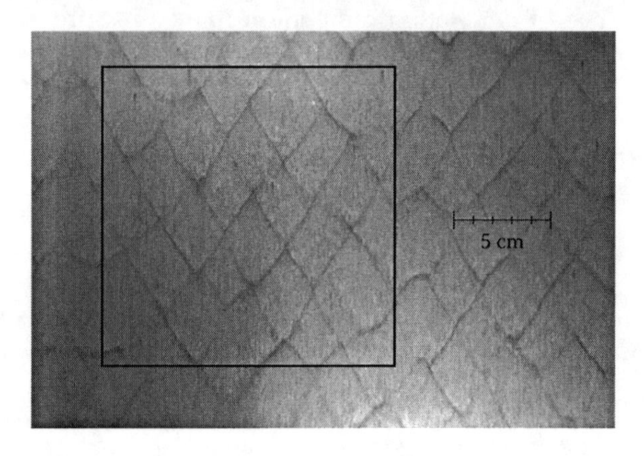

FIGURE 4.11
An actual soot foil photograph from foils aligned with the axis of the tube, which is used to record the detonation cell size with characteristic dimensions usually ranging from some millimeters to some centimeters. (Photo courtesy of Explosion Dynamics Laboratory, California Institute of Technology, http://www.galcit.caltech.edu/EDL/CellImageProcessing/cellsize.html.)

The cell size is valuable in the prediction of the onset of detonation and is related to key parameters with regard to hazardous situations [35, 37]. Cell lengths for stoichiometric hydrogen-air and hydrogen-oxygen mixtures at 101.3 kPa (1 atm) are 15.9 mm and 0.6 mm, respectively. Cell size decreases when pressure increases for hydrogen-air mixtures. The cell width of hydrogen-air detonations increases significantly with the concentration of diluents (e.g., carbon dioxide and water) [2].

The ratio of the length of a detonation cell (λ) to reaction zone width (δ) appears to depend on the mixture composition and initial conditions and varies in the range of about 10^2 (Figure 4.13).

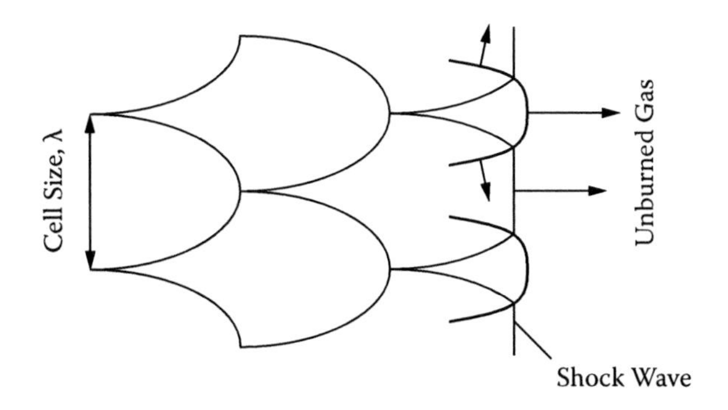

FIGURE 4.12
ZND structure and pattern of an actual structure of a detonation front. The characteristic length scale of the cell pattern, the cell size, λ, is shown in the figure. (From http://www.gexcon.com/handbook/GEXHBchap6.htm.)

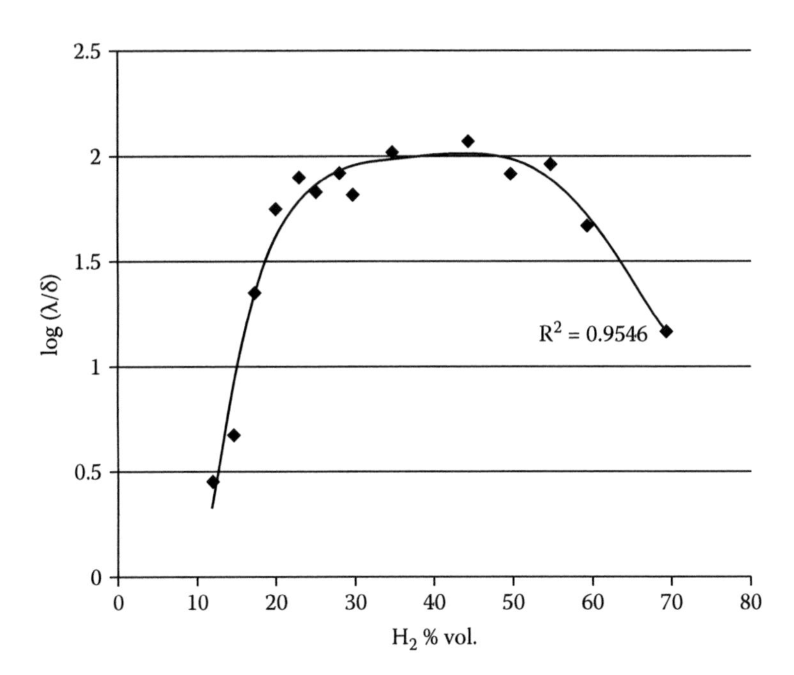

FIGURE 4.13
Dependence of the ratio of the experimental cell size (λ) to reaction zone width (δ) on H_2-air mixtures at 300 K. (Data from the Induced Chemical Reactions Laboratory, Russian Research Centre's "Kurchatov Institute," Detonation Cell Sizes found in, http://www.iacph.kiae.ru/lichr/.)

4.3.4.5 Consequences of Hydrogen-Air Explosions

With regard to the consequences of an accidental release and subsequent explosion, the energy of explosion is often taken into consideration. Calculations such as 1 gram stoichiometric hydrogen in NTP air is equivalent to 24 grams of TNT are usually found in the literature, but they are misleading since they do not take into account the weight of the oxidant (air in this case). Including air in the calculations it becomes clear that a stoichiometric hydrogen-air mixture produces only two-thirds of the explosion energy of TNT. This difference can be easily explained by the dilution effect of nitrogen in the air. In addition, it should be noticed that in an accidental release, mixing of hydrogen with air is not perfect—so only a small fraction of the theoretical energy will be produced.

Other factors that determine the catastrophic effects of an explosion are the initial density of the explosive (which is three orders of magnitude higher for TNT than for hydrogen-air mixtures) and the detonation velocity (which is three to five times higher for TNT than for hydrogen). So the resulting pressure wave from a hydrogen explosion is considerably flatter (longer duration and lower maximum overpressure) than for TNT, and the destruction effects are caused mainly by impulse rather than overpressure.

Considering the effects of a deflagration, the parameters that are usually taken into account for designing deflagration suppression systems with the aid of a suitable suppressant are maximum pressure, deflagration index, fundamental burning velocity, and effective burning velocity [28].

Maximum pressure, P_{max}, is the maximum pressure developed by a confined deflagration as determined by measurements over a wide range of concentrations.

Deflagration index, K (bar·m/s), defines the maximum rate of pressure rise with time, $(dP/dt)_{max}$, of a deflagration in a volume V, according to the equation:

$$K = (dP/dt)_{max} \times V^{1/3} \tag{4.6}$$

K_G is used for gases and K_{St} for dusts.

The rate of pressure rise is slower for larger volumes, thus allowing more time for explosion mitigation than for smaller volumes. Deflagration index measurements are undertaken in spheres of volumes of 20 L or 1 m^3 (or greater) and the results scaled up to the process vessel's volume according to the above cube-root scaling relationship.

Hazard classes (ST) have been determined for dusts based on the experimental K_{St} values as follows:

ST 0: $K_{St} = 0$ bar·m/s

ST 1: $K_{St} = 1$ to 200 bar·m/s

ST 2: $K_{St} = 201$ to 300 bar·m/s

ST 3: $K_{St} > 300$ bar·m/s

TABLE 4.14

Deflagration Maximum Pressures and Indices for Some Gases and Dusts

Gas or Dust	P_{max} (bar)	K_G or K_{St} (bar·m/s)
Methane	8.4	58
Coal dusts	8	70
Propane	8.3	103
Cellulose dusts	8	140
Hydrogen	8.2	503
Aluminum dusts	13	1,000

Source: Adapted from *Explosion Protection Specification*, Kidde Fire Protection, Oxfordshire, U.K. http://www.kfp.co.uk/utcfs/ws-438/Assets/IEP%20Explosion%20Protection %20Data%20Sheet.PDF.

However, a similar classification system has not yet been developed for vapors and gases and only limited results are given for gases. For comparison, typical P_{max} and deflagration indices for dusts and gases in larger test apparatus are given in Table 4.14 [38]. As shown in this table, hydrogen is classified as more hazardous compared to other gases, but it is much less hazardous than aluminum dust.

Deflagration suppression systems for gas and vapor explosions are designed for either quiescent or turbulent mixture conditions. In fact, the effective turbulent burning velocity is typically at least four times the value of the laminar burning velocity of the same mixture.

4.3.4.6 Damage of Structures from Thermal Radiation and Blast Waves

Exposure of structures and equipment to radiant heat flux or direct flames may be destructive, as shown in Table 4.15. With regard to assessments based

TABLE 4.15

Damage to Structures and Equipment from Thermal Radiation

Thermal Radiation Intensity (kW/m²)	Type of Damage
4	Glass breakage (30 min exposure)
12.5–15	Piloted ignition of wood, melting of plastics (>30 min exposure)
18–20	Cable insulation degrades (>30 min exposure)
10 or 20	Ignition of fuel oil (120 or 40 s, respectively)
25–32	Unpiloted ignition of wood, steel deformation (>30 min exposure)
35–37.5	Process equipment and structural damage (including storage tanks) (>30 min exposure)
100	Steel structure collapse (>30 min exposure)

Source: LaChance, J., Tchouvelev, A., and Engebo, A., *International Journal of Hydrogen Energy*, 36, 2381, 2011. With permission from Elsevier.

TABLE 4.16

Damage to Structures and Equipment from Overpressure Events

Overpressure (kPa)	Type of Damage
1	Threshold for glass breakage
15–20	Collapse of unreinforced concrete or cinderblock walls
20–30	Collapse of industrial steel frame structure
35–40	Displacement of pipe bridge; breakage of piping
70	Total destruction of buildings; heavy machinery damaged
50–100	Displacement of cylindrical storage tank; failure of pipes

Source: LaChance, J., Tchouvelev, A., and Engebo, A., *International Journal of Hydrogen Energy*, 36, 2381, 2011. With permission from Elsevier.

on a QRA approach, no probit functions are currently available in the literature relating thermal effects on structures and equipment, as is the case for effects on humans. Thus, only simplified data can be used in this case, such as those shown in Table 4.15. Nevertheless, since the exposure duration needed for thermal radiation to be effective is rather long, the impact of thermal radiation from hydrogen on structures and equipment is not considered significant.

Some damage to structures and equipment from overpressure events is shown in Table 4.16. Probit functions relating peak overpressure and the impulse of shock waves with minor, major, and total damage or collapse of structures can be found in the literature to be used in QRAs [5].

4.3.4.7 Safety Distances from Hydrogen Storage Facilities

A safety distance may be defined as the acceptable minimum separation distance between a hazard source, such as a flammable gas leak, and the object to be protected (e.g., humans or buildings). This separation distance is a function of the quantity of hydrogen and this is why we usually refer to a quantity-distance relationship. The consequences on the object and the safety distance depend on the specified threshold values of physically defined criteria, such as the dose of thermal radiation, the toxic dose, or the peak overpressure of a blast wave.

Safety distances for GH_2 and LH_2 with regard to design and operation are regulated in the United States by OSHA (the Occupational Safety and Health Administration) as part of 29 CFR (Code of Federal Regulations). According to these regulations the safety distance between a hydrogen installation and people or property is defined as 15.2 m for GH_2 in amounts more than 425 Nm^3. For LH_2 cryotanks containing more than 2.27 m^3, the safety distance from the same objects is at least 22.8 m [39]. In Europe, LH_2 storage tanks are covered by the recommendations given by EIGA (European Industrial Gases Association) specifying minimum safety distances [40]. For instance, the specified minimum safety distance of LH_2 storage tanks from combustible liquids or solids, roads, railroads, overhead power lines, and technical buildings is 10 m. (For more details consult Chapter 13.)

4.3.4.8 Deflagration-to-Detonation Transition

As previously discussed, the oxidation (combustion) of hydrogen in air or other oxidant (e.g., pure oxygen) can be subsonic (deflagration) or supersonic (detonation) depending, among other factors, on the strength of the ignition source. A highly energetic source such as a shockwave can initiate a detonation, whereas a low strength energy source such as a flame or a spark will typically result in a deflagration.

Nevertheless, there is a possibility that a deflagration will transition to a detonation after the flame has traveled for some time and distance by the following mechanism. Although a deflagration wave is subsonic, it can perturb the gas ahead of the flame front by generated compression waves due to the expansion of combustion products. A newly produced compression wave propagates at a higher velocity through the gas which is pre-compressed and adiabatically heated by the previous compression waves. Thus, a shock wave will be formed and strengthened when overtaken by successive compression waves and finally transformed to a shock wave. At the moment the strength of the precursor shock reaches a critical value, a sudden change of the propagation mechanism occurs from the diffusion and heat conduction controlled subsonic propagation to adiabatic shock wave driven propagation. This *deflagration-to-detonation transition* (DDT) is not a continuous process but rather a stepwise change [41, 42].

In general, the composition range in which a detonation can take place is narrower than that of a deflagration. The range for detonation limits found in Table 4.10 for hydrogen-air mixtures is from 18.3 to 59 percent hydrogen. However, with sufficiently strong ignition sources, these limits can be extended and, in addition, an overdriven detonation can be obtained; this results in a higher detonation velocity which gradually declines to a stable detonation at a thermodynamically determined lower velocity. Even in the absence of a strong initiator, a deflagration can be accelerated to detonation when other factors, such as sufficient degree of confinement and the presence of obstacles inducing turbulence in the flame front, limit energy losses and suitably mix the ingredients, thus triggering a detonation. The abrupt increase of overpressure and flame speed on the transition from deflagration to detonation for hydrogen-air mixtures has been modeled by Eichert, whose work was based on numerical calculations of 1-d tube combustion processes [42].

4.3.4.9 Accidental Hydrogen Generation in Nuclear Reactors

Following severe accidents in nuclear power plants, substantial amounts of hydrogen can be generated, for instance, from the chemical reaction between zirconium cladding and hot water vapor, as well as from core-concrete interactions after a lower head failure of the vessel. Hydrogen thus generated may find its way into the compartments in the containment building with the potential to threaten the containment integrity by over-pressurization from deflagrations or detonations. Moreover, even local hydrogen burning, which is not a threat to the global containment integrity, may also be a hazard to the survivability of safety-related equipment.

Hence, regulations for hydrogen control based on local hydrogen concentrations have been established in countries using nuclear reactors. They specify that the average hydrogen concentration in the containment to be controlled is not to exceed 10 percent during and following an accident that releases an equivalent amount of hydrogen which would be generated from a 100 percent fuel clad metal-water interaction, or that the post-accident atmosphere will not support hydrogen combustion. Therefore, equipment with hydrogen control units such as igniters or catalytic recombiners has been considered [43].

The mechanism by which hydrogen is produced in nuclear power plants is the following. At high temperatures, zircaloy, which is used as a metal cladding in nuclear reactors, can react with water to produce hydrogen. This exothermic reaction followed by a deflagration or detonation of the hydrogen-air mixture can produce chemical power to accelerate the progression of a severe accident. The zircaloy–water reaction, which turns to a run-away reaction at temperatures greater than about 1200°C, is as follows:

$$Zr + 2H_2O \rightarrow ZrO_2 + 2H_2 + \Delta H_{Zr}$$

where ΔH_{Zr} is the heat of the reaction per mole of zircaloy consumed, currently set at 616 MJ/kmol. This oxidation is assumed to take place at the Zr/ZrO_2 interface leading to an increase in the oxide layer thickness.

One effective method to reduce the containment hydrogen concentration in these cases is to intentionally burn the hydrogen generated with igniters to avoid severe consequences from hydrogen-air explosions (Figure 4.14). Continuous burning may prevent hydrogen accumulation and not result in a significant pressure spike in the containment building. Once pressure and temperature spikes are over, no more short-term challenges to containment integrity remain. On the basis of a best-estimate assessment, it has been shown that it is beneficial to utilize hydrogen igniters rather than do nothing with respect to the expected value of hydrogen concentration in the containment building during an accident [44].

4.3.5 Environmental Concerns

Although hydrogen is considered environmentally friendly compared to hydrocarbons as a transportation fuel, a study by researchers from the California Institute of Technology (CalTech) has shown that if we pass to the so-called hydrogen economy, the hydrogen leakage from its extended use

FIGURE 4.14 (See color insert.)
A photo from video footage taken March 14, 2011, shows the hydrogen explosion at the Fukushima Daiichi nuclear power complex at the moment of the event and afterward. (Reuters/NTV photo via Reuters TV. With permission.)

will be as much as 10 to 20 percent [45]. This would result in the rapid escape of huge quantities (estimated as 60 to 120 thousand tonnes) of this extremely light gas to the ozone layer, resulting in doubling or tripling of hydrogen input into the atmosphere from all current natural or human sources. The output will then be the creation of additional water which would cool and dampen the stratosphere, finally thinning the stratospheric ozone layer by as much as 10 percent. On the other hand, the combination of hydrogen with oxygen to form water would create increased noctilucent clouds (high wispy tendrils) appearing at dawn and dusk which would accelerate global warming.

If the CalTech study is finally verified, hydrogen impact in the environment would resemble the catastrophic effect chlorofluorocarbons (CFCs) had on the stratospheric ozone layer, as evidenced by the annual holes over the South and North poles. Since no one would like to repeat the errors of the past, there is still time to fully investigate this eventuality and develop cost-effective technologies to minimize leakage before entering a global hydrogen economy.

Nevertheless, there still remain many uncertainties about the hydrogen cycle in the atmosphere. Moreover, the anticipated accumulation of hydrogen in the air is questioned by other scientists and organizations, such as the U.S. Department of Energy (Office of Energy Efficiency and Renewable Energy) and the U.S. National Hydrogen Association, claiming that the increase in the total hydrogen concentration would be at least one order of magnitude less than the CalTech researchers estimate. This would result in less than a 1 percent increase in ozone depletion considering the worst case scenario.

References

1. Fact Sheet Series No. 1.008, "Hydrogen Safety," National Hydrogen Association, Washington, D.C.
2. ANSI, *Guide to Safety of Hydrogen and Hydrogen Systems,* American Institute of Aeronautics and Astronautics, American National Standard ANSI/AIAA G-095-2004, 2004, Chap. 2.
3. Zuettel, A., Borgschulte, A., and Schlapbach, L. (Eds.), *Hydrogen as a Future Energy Carrier*, Wiley-VCH Verlag, Berlin, Germany, 2008, Chap. 4.
4. MedicineNet.com, http://www.medterms.com.
5. LaChance, J., Tchouvelev, A., and Engebo, A., Development of uniform harm criteria for use in quantitative risk analysis of the hydrogen infrastructure, *International Journal of Hydrogen Energy,* 36, 2381, 2011.
6. Methods for the determination of possible damage. In CPR 16E. The Netherlands Organization of Applied Scientific Research, 1989.
7. Eisenberg, N.A., et al. *Vulnerability Model: A Simulation System for Assessing Damage Resulting from Marine Spills,* Final Report SA/A-015 245, U.S. Coast Guard, 1975.
8. Tsao, C.K., and Perry, W.W., *Modifications to the Vulnerability Model: A Simulation System for Assessing Damage Resulting from Marine Spills,* Report ADA 075 231, U.S. Coast Guard, 1979.
9. Opschoor, G., van Loo, R.O.M., and Pasman, H.J., Methods for calculation of damage resulting from physical effects of the accidental release of dangerous materials. International Conference on Hazard Identification and Risk Analysis, Human Factors, and Human Reliability in Process Safety. Orlando, Florida, January 15–17, 1992.
10. Lees, F.P., The assessment of major hazards: a model for fatal injury from burns, *Transactions of the Institution of Chemical Engineers,* 72 (Part B), 127–134, 1994.
11. Jeffries, R.M., Hunt, S.J., and Gould, L., Derivation of probability of fatality function for occupant buildings subject to blast loads, *Health & Safety Executive,* Contract Research Report 147, 1997.
12. *Guidelines for Use of Hydrogen Fuel in Commercial Vehicles,* DOT F 1700.7, Report No. FMCSA-RRT-07-020, U.S. Department of Transportation, Washington, D.C., 2007.

13. ANSI, *Guide to Safety of Hydrogen and Hydrogen Systems,* American Institute of Aeronautics and Astronautics, American National Standard ANSI/AIAA G-095-2004, Chap. 3.
14. Chandler, W.T., and Walter, R.J., Testing to determine the effect of high pressure hydrogen environments on the mechanical properties of metals, in *Hydrogen Embrittlement Testing,* ASTM 543, American Society for Testing and Materials, Philadelphia, 1974, 170.
15. Groenvald, T. D. and Elcea, A. D. Hydrogen Stress Cracking in Natural Gas Transmission Pipelines. *Hydrogen in Metals: Proceedings of an International Conference on the Effects of Hydrogen on Materials Properties and Selection of Structural Design,* I. M. Bernstein and A. W. Thompson, Eds., ASM International, September, 1973.
16. Rowe, M.D., Nelson, T.W., and Lippold, J.C., Hydrogen-induced cracking along the fusion boundary of dissimilar metal welds, *Welding Research,* February 1999, 31.
17. Zalosh, R.G., and Short, T.P., *Compilation and Analysis of Hydrogen Accident Reports,* C00-4442-4, Department of Labor, Occupational Safety and Health Administration, Factory Mutual Research Corp., Norwood, 1978.
18. Williams, G.P., Causes of ammonia plant shutdowns, *Chemical Engineering Progress,* 74, 9, 1978.
19. Ordin, P.M., A review of hydrogen accidents and incidents in NASA operations, in *9th Intersociety Energy Conversion Engineering Conference,* American Society of Mechanical Engineers, New York, 1974.
20. Thompson, J.D., and Enloe, J.D., Flammability limits of hydrogen-oxygen-nitrogen mixtures at low pressures, *Combustion and Flame,* 10(4), 393–394, 1996.
21. Fisher, M., Safety aspects of hydrogen combustion in hydrogen energy systems, *International Journal of Hydrogen Energy,* 11(9), 593–601, 1986.
22. McHale, E.T., Geary, G., von Elbe, G., and Huggett, C., Flammability limits of H_2-O_2-fluorocarbon mixtures, *Combustion and Flame,* 16, 167, 1971.
23. Creitz, E.C., Inhibition of diffusion flames by methyl bromide and trifluoromethyl-bromide applied to the fuel and oxygen sides of the reaction zone. *Journal of Research for Applied Physics and Chemistry,* 65, 389, 1961.
24. Wionsky, S.G., Predicting flammable material classifications, *Chemical Engineering,* 79, 81, 1972.
25. Beach, R., Preventing static electricity fires, *Chem. Eng.,* 71, 73, 1964.
26. Beach, R., Preventing static electricity fires, *Chem. Eng.,* 72, 63, 1965a.
27. Beach, R., Preventing static electricity fires, *Chem. Eng.,* 72, 85, 1965b.
28. FM Approval CN 5700 – Approval Standard for Explosion Suppression Systems, FM Approvals, Norwood, Massachusetts, 2002.
29. Baker, W.E., and Tang, M.J., *Gas, Dust and Hybrid Explosions,* Elsevier, Amsterdam, 1991, 42.
30. Center for Chemical Process Safety, *Guidelines for Evaluating the Characteristics of Vapor Cloud Explosions, Flash Fires, and BLEVEs,* American Institute of Chemical Engineers, New York, 1994.
31. Lide, D.R., Ed., *Handbook of Chemistry and Physics,* 75th ed., CRC Press, Boca Raton, FL, Chap. 6, 1994.
32. Salla, J.M., Demichela, M., and Casal, J., BLEVE: A new approach to the superheat limit temperature, *Journal of Loss Prevention in the Process Industries,* 19, 690, 2006.

33. Bjerketvedt, D., Bakke, J.R., and Wingerden, K.V., Gas explosion handbook. *Journal of Hazardous Material*, 52, 1, 1997.
34. Philips, H., *Explosions in the Process Industries*, Institute of Chemical Engineers, Warwickshire, U.K., 1994, 5.
35. Lee, J. H. et al., Hydrogen-air detonations, in *Proc. 2nd International Workshop on the Impact of Hydrogen on Water Reactor Safety*, M. Berman, Ed., SAND82-2456, Sandia National Laboratories, Albuquerque, NM, 1982.
36. Gordon, S., and McBride, B.J., *Computer Program for Calculation of Complex Chemical Equilibrium Compositions and Applications*, NASA Reference Publication 1311, 1994.
37. Bull, D.C., Ellworth, J.E., and Shiff, P.J., Detonation cell structures in fuel/air mixtures, *Combustion and Flame*, 45, 7, 1982.
38. *Explosion Protection Specification*, Kidde Fire Protection, Oxfordshire, U.K. http://www.kfp.co.uk/utcfs/ws-438/Assets/IEP%20Explosion%20Protection%20Data%20Sheet.PDF.
39. U.S. DOT, Clean Air Programme – Use of Hydrogen to Power the Advanced Technology Transit Bus (ATTB): An Assessment. Report DOT-FTA-MA-26-0001-97-1, U.S. Department of Transportation, Washington D.C., 1997.
40. European Industrial Gases Association, *Safety in Storage. Handling and Distribution of Liquid Hydrogen*, Report DOC 06/02/E, 2002.
41. Lee, J., Initiation of gaseous detonation, *Annual Review of Physical Chemistry*, 28, 75, 1977.
42. Eichert, H., Hydrogen-air deflagrations and detonations: numerical calculation of 1-d tube combustion processes, *International Journal of Hydrogen Energy*, 12, 171, 1987.
43. Choi, Y.S., Lee, U.J., Lee, J.J., and Park, G.C., Improvement of HYCA3D code and experimental verification in rectangular geometry, *Nuclear Engineering and Design*, 226, 337, 2003.
44. Lee, S.D., Suha, K.Y., and Jae, M., A framework for evaluating hydrogen control and management, *Reliability Engineering and System Safety*, 82, 307, 2003.
45. Tromp, T.K., Shia, R.L., Allen, M., Eiler, J.M., and Yung, Y.L., Potential environmental impact of a hydrogen economy on the stratosphere, *Science*, June 13, 2003, 1740–1742. DOI: 10.1126/science.1085169.

5

Hazards in Hydrogen Storage Facilities

5.1 Storage Options

The storage options for hydrogen (including those under investigation) are the following [1]:

- Compressed hydrogen gas (GH_2) in cylinders or tanks
- Tethered balloons, "bags," or water displacement tanks (low-pressure GH_2)
- Hydrogen adsorbed into metal to form a metal hydride (MH)
- Liquid hydrogen (LH_2) in cryogenic tanks
- Adsorption on high-surface-area carbon powder in tanks
- Encapsulation in glass microspheres (experimental)
- Adsorption on carbon nanotubes (experimental)
- In water (H_2O; not a "fuel")
- In ammonia (NH_3)
- In liquid hydrocarbons: gasoline, diesel fuel, alcohol, liquid natural gas (LNG), propane or butane (LPG), etc.
- In gaseous hydrocarbons: compressed natural gas (CNG), biogas, etc.

The first three hydrogen storage options above, namely, GH_2, MH, and LH_2, are the "state of-the-art" methods that are the most frequently applied in vehicular and stationary applications. The last two hydrogen storage options are the fossil fuels (liquid and gaseous hydrocarbons) that currently dominate our global fuel production and consumption.

Liquid hydrocarbons are widely used in transportation because of their extremely high *energy density* (Table 5.1). The values of properties in this table were determined with the Computational Knowledge Engine WolframAlpha [2]. It is interesting to note that although the gravimetric energy density of LH_2 is three times greater than that of gasoline, its volumetric energy density is less than three times lower than gasoline's. Furthermore, liquid hydrogen

TABLE 5.1

Hydrogen Densities in Different Forms of Hydrogen Storage

Fuel	Formula	Density* (kg/m³)	Gravimetric Energy Density* (MJ/kg)	Volumetric Energy Density (MJ/L)	H_2 Density (kg H_2/m³)
GH_2	H_2	0.09	142	0.013	0.09
LH_2	H_2	71	142	10.2	71
LNG (methane)	CH_4	424	55.5	23.5	106
LPG (propane)	C_3H_8	582	50.1	29.2	106
Gasoline	C_8H_{18}	737	47.3	34.9	118
Methanol	CH_3OH	791	22.7	18	99
Ethanol	C_2H_5OH	789	29.7	23.4	103
Cyclohexane	C_6H_{12}	779	46.7	36.4	111
Methylcyclohexane	C_7H_{14}	770	46.6	35.9	95
Ammonia (liquid at BP)	NH_3	683	18.6	12.7	121
Hydrazine	N_2H_4	1011	19.2	19.4	126
Water	H_2O	1000	—	—	111

* WolframAlpha Computational Knowledge Engine was used for the determination of property values (http://www.wolframalpha.com).

has much less H_2 density in kilograms per cubic meter than other chemical hydrogen storage media that are currently used or are intended to be used as energy carriers. Among them, ammonia and hydrazine possess the greatest H_2 densities, but their use in common applications is not possible because of their high chemical reactivity and toxicity. Nevertheless, hydrazine has been used effectively since World War II as a rocket fuel or monopropellant in space exploration (e.g., *Viking* and *Phoenix* landers), in military aircraft (e.g., F-16 Fighter aircraft), and other industrial uses (e.g., in polymers and pharmaceuticals and as a gas precursor in air bags).

However, water possessing very high H_2 density is the hydrogen storage reservoir most likely to be used in the future after the development of reasonably priced methods for hydrogen separation from water.

5.1.1 Storage as Liquid Hydrogen

There are cases where liquid hydrogen is beneficial, for instance, when high purity is needed. Disadvantages are boil-off losses, the temperature stability necessary to avoid overpressure, and, mainly, the liquefaction energy requirement. Technology for storage of LH_2 is commercially available in vessels with sizes ranging from 0.1 to thousands of cubic meters. The main concern in LH_2 tank design is the construction of an effectively insulated container. This is normally obtained with vacuum-jacketed, double-walled containers. The preferred shape is spherical due to its minimum surface-to-volume ratio and its more uniform distribution of stresses and strains [3].

Special attention is given to the multilayer construction of the insulation, which may consist of 60 to 100 reflective foils fixed on the outside of the inner vessel, with spacers between each layer acting as thermal barriers and a total thickness of at least 20 mm for tank sizes up to 300 m^3. The void volume between the two vessel walls may be filled with reflective perlite powder or hollow glass microspheres that considerably reduce heat loss and allow for less demanding vacuum than normally needed (1.3 Pa). For safety reasons care is taken with regard to the potential shifting of insulation particles during contraction of the inner vessel, which leads to compaction upon re-expansion that may rupture the supporting structures. Suitable materials for construction of cryogenic tanks are carbon steel for the outer vessel and stainless steel or aluminum for the inner vessel. Tubing is generally made from stainless steel [3, 4].

Except reflective materials, liquid nitrogen (LN_2) is usually used for safety reasons in large LH_2 tanks to fill the space created by an additional outer wall. Up to now, NASA has constructed the largest LH_2 tanks, including two identical storage tanks of 3,800 m^3 capacity each, at the Kennedy Space Center in Florida, for its space shuttle program. The construction materials are austenitic stainless steel for the inner wall, with an inside diameter of 18.75 m, and carbon steel for the outer wall, with an inside diameter of 21.34 m. The ullage is about 15 percent, permitting each tank to be filled up to 3,218 m^3. Operation pressure is 620 kPa, with a boil-off rate equal to 0.025 percent per day, or about 800 L per day.

5.1.2 Storage in Porous Media

Increased safety accompanies storage of hydrogen in porous media, as well as lower pressure storage and design flexibility, but this technology is not ready to be used yet. Reversible sorption in porous media may be physisorpion (van der Waals forces) or chemisorption (as in metal hydrides). The materials that are extensively studied as sorbents are [4]:

- Carbon-based materials, nanotubes, nanofibers, activated carbons, activated fibers, carbons from templates, powders, doped carbons, and cubic boron nitride alloys
- Organics, polymers, zeolites, silicas (aerogels), porous silicon

Comparing various types of *storage in carbon-based materials,* nanotubes appear to have higher storage capacity than activated carbons. Yet, research data for nanotubes are sometimes conflicting due probably to uncertainties in the material used.

Storage in other noncarbon materials includes:

- Self-assembled nanocomposites/aerosols. These are nanostructured open-celled foams with very low densities, inexpensive, and safe to be used for hydrogen storage by physisorption.
- Zeolites. These are crystalline nanoporous materials, available at low cost, environmentally friendly, and safe to be used.
- Metal organic materials. These are usually zeolitic-type materials with backbone made of carbon possessing tailored properties and high potential.
- Other materials, such as glass microspheres, hydride slurries, boron nitride nanotubes, bulk amorphous materials (BAMs), hydrogenated amorphous carbon, hybrids, metal-organic frameworks (MOFs), and sodium borohydride.

5.2 Hazard Spotting

In atmospheric conditions, hydrogen is a colorless, odorless gas much lighter than air. The low density in conjunction with the small particle size allows the penetration of hydrogen molecules in some metals and alloys such as cast-iron and high-carbon steel [5]. The penetration may end in small hydrogen leaks, or, in the presence of cracks within the wall, contribute to crack spread, material strength decrease, and subsequent fracture.

Hydrogen reacts violently with oxidizers such as nitrous oxide, halogens (especially with fluorine and chlorine), and unsaturated hydrocarbons (i.e., acetylene) with intensely exothermic reactions. Hydrogen gas forms combustible or explosive mixtures with atmospheric oxygen over a wide range of concentrations in the air in the range 4.0–75 percent v/v and 18–59 percent v/v, respectively. Since hydrogen has much broader flammability and detonability limits than any other fuel, it should never be stored unless it is below the lower flammability limit (LFL). Thus, industry standards for storage safety are set well below 0.25 of LFL, that is, less than 1 percent oxygen with the hydrogen [1].

Jet fires are visually not detectable due to the invisibility of hydrogen flames, especially in daylight, so a hydrogen fire is very difficult to spot and this has resulted in delayed action and serious injuries of carelessly approaching persons.

Hydrogen is nontoxic; however, besides the hazards of flash fire, jet fire, and gas cloud explosion in the event of an accidental release, hydrogen may act as an asphyxiant at sufficiently high concentrations by depleting the oxygen available in atmospheric air.

5.2.1 Refrigerated Storage Conditions

Hydrogen is stored at normal temperature either in a gaseous state under medium (4.1 to 8.6 bar) or high pressure (140 to 400 bar), or in the liquid state under low temperature and moderate pressure. When stored at medium pressure, hydrogen should be kept in low-carbon steel or other materials unaffected by hydrogen embrittlement. High-carbon steel tanks are not suitable for storing hydrogen under pressure. For preventing hydrogen tank embrittlement, cold-rolled or cold-forged steels and those having weld hard spots in excess of about Vickers Hardness Number 260 should be avoided. Nonmetal tanks such as composite-fiber tanks avoid hydrogen embrittlement concerns and derating.

Medium-pressure GH_2 storage tanks have usually smaller size and greater weight, for a given storage capacity, relative to low-pressure GH_2 tanks. Storage tanks for hydrogen should be hydrostatically tested to at least twice the operating pressure, equipped with a pressure release valve, and always installed outdoors for safety reasons. In addition, inlet and outlet lines should be equipped with flash-back arrestors [1].

In any case there is a significant hazard potential for mechanical explosion of hydrogen vessels (tanks and cylinders) when exposed to high temperature or thermal radiation. Usually, the cause of overheating is a neighboring fire (primary event) that causes temperature increase in the vessel shell and the content (Figure 5.1). The vessel finally bursts (secondary

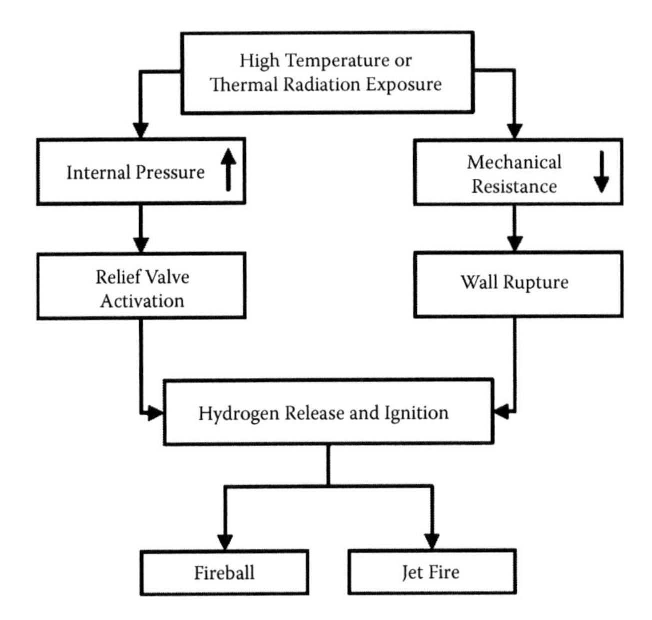

FIGURE 5.1
Thermal radiation effect on pressurized hydrogen vessels. (From Rigas, F., and Sklavounos, S., *International Journal of Hydrogen Energy*, 30, 1501, 2005. With permission from Elsevier.)

event) spilling its content, which, if flammable, is usually immediately ignited and burns in the form of a jet fire or a fireball. Typical ignition sources for the spilling fuel may be electrostatic sparks developed during content discharging or the flames of the neighboring fire. This situation is known as a domino effect since an initial accident gives rise to another, thus generating a chain of accidents with escalating impact on the surroundings [6, 7].

In contrast to storage vessels for high-pressure gaseous hydrogen (GH_2), liquefied hydrogen vessels operate at moderate to low pressures that normally do not exceed 20 bar. As a result, it is reasonable for the walls to be designed with lower pressure resistance than that for storage of hydrogen gas (which may have operation pressures up to 400 bar, thus being liable to increased risk of failure). In the event that the container is engulfed in a fire, its metal is heated and loses mechanical strength. While the liquid phase absorbs significant amounts of heat, vapors possess far lower specific heat capacity. Therefore, the heat supplied to that part of the container where the vapor phase exists will raise the local wall temperature much more, thus weakening its strength [6].

Concerning liquefied gas storage, vessel overheating may result in internal temperatures higher than the boiling point of the content, indeed without vaporization initiation of the liquid phase, and then the liquid is superheated. This phenomenon is observed when there is a shortage of nucleation sites (i.e., impurities, crystals, or ions) in the bulk of the liquid. However, there is a temperature limit above which the fluid cannot remain in the liquid state (the homogeneous nucleation limit or superheat limit temperature). In this limit, random molecular density fluctuations within the bulk of a liquid produce hole-like regions of such molecular dimensions that may act as bubbles [8]. The end result is the explosive flash of the liquid accompanied by a strong shock wave that propagates through the fluid and ruptures the container spilling the content in the atmosphere (boiling liquid expanding vapor explosion or BLEVE; see Chapter 4) [9]. Missiles of the ruptured walls may travel hundreds of meters, whereas the flammable content ignites forming a sphere that burns from the outer to inner layers and is known as a *fireball*. The overall process is shown in Figure 5.2.

Research on modeling fireball phenomena was recently carried out by Sklavounos and Rigas [10] with reasonable quantitative estimations of the thermal load received in the vicinity. A good qualitative simulation was also accomplished for the formation and evolution of the fireball after initiation of burning (Figure 5.3). Fireball development was based on release and ignition of 1,708 kg of propane. The burning cloud moved upward as expected, obtaining positive buoyancy due to high temperatures, whereas air inflow from the left (inlet boundary condition) caused horizontal fireball shift to the right.

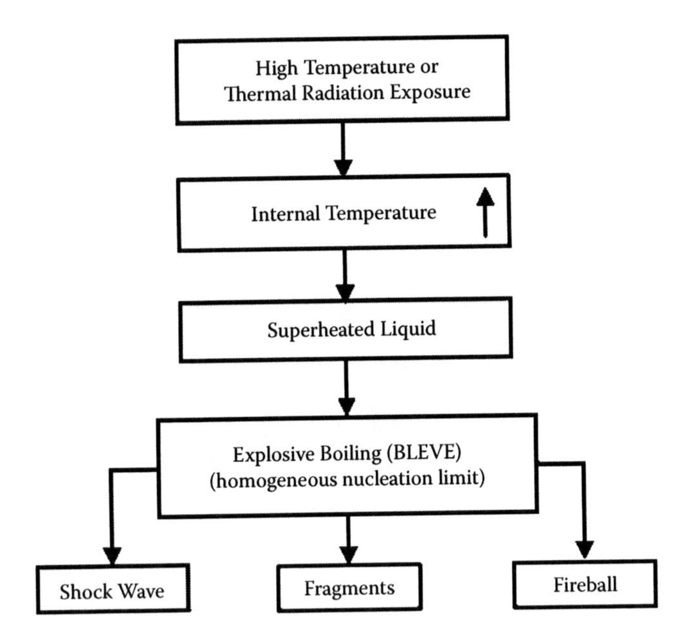

FIGURE 5.2
Thermal radiation effect on liquefied hydrogen vessel. (From Rigas, F. and Sklavounos, S., *International Journal of Hydrogen Energy*, 30, 1501, 2005. With permission from Elsevier.)

5.2.2 Cryogenic Storage Conditions

At hydrogen refilling stations and in vehicles, hydrogen has to be compressed at high pressures (up to 400 bar), due to its low energy content per unit volume. Nevertheless, in some applications, hydrogen, as well as other gases (e.g., carbon dioxide, nitrogen, helium, and methane), need to be stored in a liquid state at very low temperatures for volume restriction. These temperatures are often lower than –73°C, so that storage conditions are characterized as cryogenic (as distinguished from the refrigerated condition). These conditions are preferred, for example, in uses such as rocket propulsion, as well as in warehousing for convenience and economy. A crucial factor in liquefaction is the critical temperature above which the gas cannot be liquefied by pressure application only.

In practice, hydrogen is kept liquefied at extremely low temperatures, below –240.2°C, and moderate pressures (20–30 bar). Major hazards related to cryogenic storage stem from:

- *Embrittlement of service materials.* The low temperatures inside storage tanks and transmission pipelines may cause significant susceptibility of the structural material to vibrations and shocks. Mild steel and most iron alloys at liquid hydrogen (LH_2) temperatures lose their ductility, being liable to increased risk of mechanical failure [11]. Figure 4.1 in Chapter 4 presents the dramatic decrease of impact

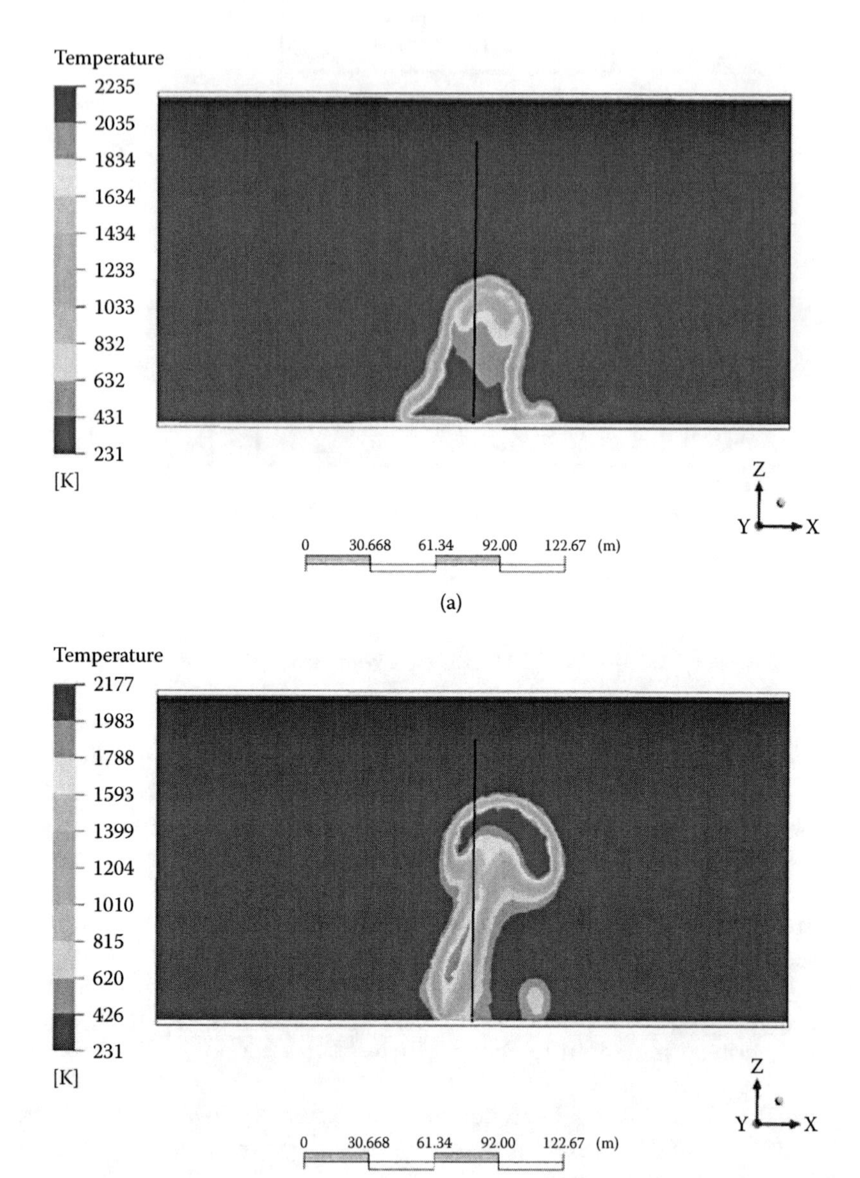

FIGURE 5.3 (See color insert.)
Fireball development after release and ignition of 1708 kg of propane. The burning cloud moves upward, obtaining positive buoyancy due to high temperatures, whereas air inflow from the left (inlet boundary condition) causes horizontal fireball shift to the right. (a) 1.2 s after ignition, (b) 3.2 s after ignition. (From Sklavounos, S., and Rigas, F., *Process Safety and Environmental Protection*, 90(2), 129, 2012. With permission from Elsevier.)

strength with a decrease in temperature. The same risk exists with other elements of equipment (e.g., control valves, gauges) incorporated in vessels and pipes.

- *Liquid hydrogen spills.* Liquefied hydrogen releases yield dramatically larger volumes of combustible clouds (1 L of liquid produces on evaporation 851 L of gas). Therefore, the consequences of a fire or explosion will be more extensive than that of a pressurized hydrogen release.
- *Extremely low temperature.* Low temperature can cause severe tissue frostbite if the material comes into contact with the human body. Flesh may adhere quickly to cold, insufficiently insulated pipes or vessels and tear on attempting to withdraw it [12].
- *Hydrogen cloud dispersion.* A hydrogen spill that originates from cryogenic storage results in cloud formation that disperses in a manner similar to a heavier-than-air gas, thus increasing the risk of accidental fires and explosions.

5.3 Hazard Evaluation

5.3.1 Methodology

Hazard evaluation in hydrogen storage facilities aims at determining all credible accident scenarios. A variety of methods (event tree, failure modes and effects analysis, what-if, fault tree) can be found in the relevant literature [13-15]. (See also Section 8.2.4.) Of these, event tree analysis (ETA) is a formal technique and one of the standard approaches used for industrial incident investigation. ETA is a logic model that graphically portrays the combination of events and circumstances in an accident sequence. It is an inductive method, which begins with an initiating undesirable event and works toward a final result (outcome). The general procedure for ETA involves the following steps:

1. Determination of the initiating events that can result in certain types of accident.
2. Identification of the critical factors that may affect the evolution of the initiating event.
3. Construction of the event tree taking into account the interaction between critical factors and the initiating event.
4. Designation and evaluation of resulting accidental events.

Applying ETA in fuel gas releases, the critical factors that may substantially affect the final outcome are the time of ignition of the resulting cloud and the confinement provided by the surroundings. The former is related to the mixing of escaping fuel gas with air. When immediate ignition occurs, gas cloud mixing with atmospheric oxygen is still limited; thus, ignition takes place on the outer layer that is between the flammable limits, whereas the inner core of the cloud is too rich in fuel to ignite. As buoyancy forces of the hot gases begin to dominate, the burning cloud rises and becomes more spherical in shape forming a ball of flames. This elevation gradually causes further mixing of the gas with oxygen, which brings new volumes of gas into the flammable limits, thus sustaining the fire. In contrast, when delayed ignition occurs, the fuel cloud may have been adequately mixed with air, so that after ignition it flashes back. This differs from a fireball scenario since it proceeds faster and can burn from inner to outer flammable layers provided a proper ignition source is found there. Thus, a deflagration or detonation may occur, the latter necessitating a more uniform (and in narrower concentration limits) mixing of hydrogen with air, and in addition some increased degree of confinement.

5.3.2 ETA Method Application

Rigas and Sklavounos [7] have applied the ETA method with regard to accidental hydrogen release as shown in Figure 5.4. It is obvious from this figure, that, unless an immediate ignition takes place, there is some time of dispersion that intervenes between release and ignition. Generally, if hydrogen flammability zones were known, it would be possible to take preventive measures and to prepare emergency response planning against fires and explosions.

Consequently, a major issue arises regarding the computation of the dispersion succeeding an accidental hydrogen release. Furthermore, even if no ignition takes place, escaping hydrogen may accumulate into closed spaces adjacent to the source posing an asphyxiation hazard for the people there. Hydrogen dispersion may be considered safe only when no ignition occurs and no confined space exists [3].

In this section, the main hazards associated with hydrogen storage procedures were analyzed. In addition, possible accidental events that hydrogen may yield were indicated by performing an event tree analysis of a hydrogen release. The analysis showed that, unless an immediate ignition occurs, the determination of the low flammability limit (LFL) distance of the fuel-air mixture is of major importance for the purpose of loss prevention, such as ignition source elimination.

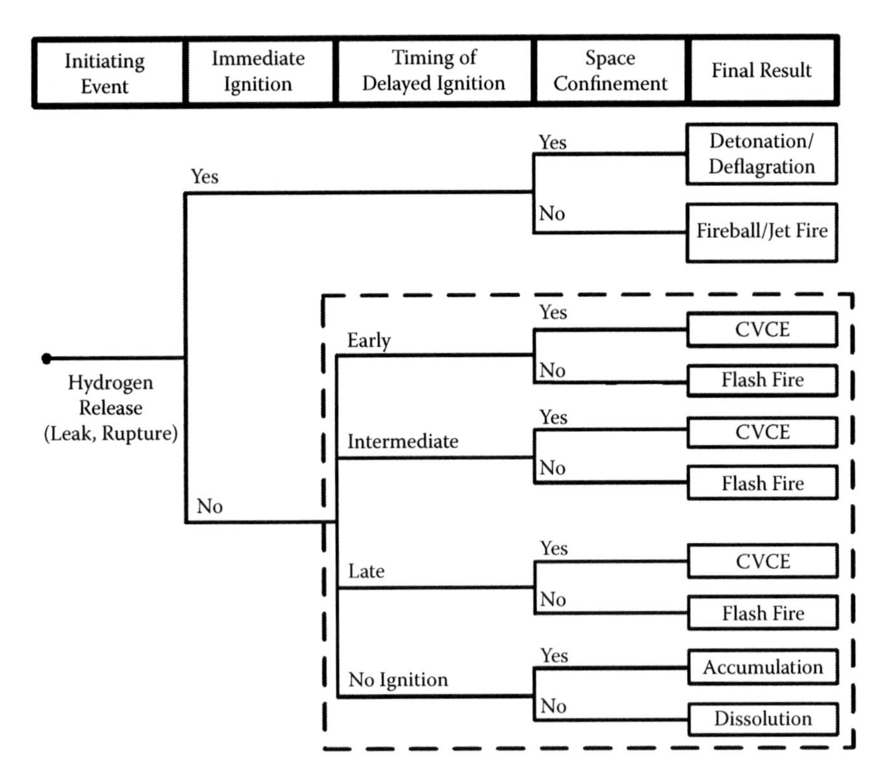

FIGURE 5.4
Event tree analysis adapted to accidental hydrogen releases. (From Rigas, F., and Sklavounos, S., *International Journal of Hydrogen Energy*, 30, 1501, 2005. With permission from Elsevier.)

5.4 Qualitative Prediction of Cloud Travel

Generally, the behavior of a gas (or vapor) cloud during dispersion can be either buoyant or gravity driven. The former is relevant with lighter-than-air gases and the latter with heavier-than-air gases.

Currently, dispersion estimations are performed via the use of semiempirical, one-dimensional models (the so-called box models) and computational fluid dynamics (CFD) codes. The selection of the appropriate dispersion model in an accidental release scenario requires the behavior of the dispersing gas to be known since each model is specialized for one type of release (buoyancy or gravity driven).

At atmospheric conditions (20°C, 1 atm), hydrogen is a lighter-than-air gas and preserves the property of light gas behavior even when releases from high-pressure storage systems occur. However, once hydrogen escapes while stored in a liquid state, its dispersion behavior turns to that of a heavy rather than a light gas.

The different way in which a gas disperses according to the storage conditions has been observed in field-scale trials at Lawrence Livermore National Laboratory (LLNL) installations with another light gas (natural gas) that was left to escape and disperse while stored in a liquid state [16]. In this case, instead of rising the vapor cloud moved along the ground, being under the LFL limit, for several tens of meters.

5.5 Gas Dispersion Simulation

5.5.1 CFD Modeling

Generally, CFD methods follow a deterministic procedure to approximate a problem. They consider the fundamental Navier-Stokes equations for mass, momentum, and heat transfer processes, in conjunction with other partial differential equations for describing further processes, such as turbulent mixing between the dispersing gas and the atmospheric air.

The first step in a CFD simulation attempt includes the definition of the three-dimensional geometry (computational domain), in addition to the subdivision of the entire domain into a number of smaller control volumes (cells) that form a mesh. The geometry comprises a unified set of parametric surfaces built in an appropriate CAD interface.

In CFX (a CFD code), the solution process begins with discretizing the spatial domain into finite control volumes using the mesh constructed in the preprocessor stage. Integration of the subsequent balance equations over each control volume is based on the finite volume method (FVM) [17]. The integral equations obtained are converted to a system of algebraic equations using a high-resolution numerical scheme and are solved iteratively at nodal points inside each cell aiming at minimization of the residuals to acceptable convergence levels. As far as their solution is concerned, CFX uses a coupled solver that solves the hydrodynamic equations (for pressure and velocity components) as a single system. This solution approach uses a fully implicit discretization of the equations at any given time step reducing the number of iterations needed for convergence.

5.5.2 CFD Simulations of Gaseous and Liquefied Hydrogen Releases

Sklavounos and Rigas [18], working on large-scale gas dispersions, have used the general-purpose CFD code named CFX, first validating it against experimental data from large-scale gas dispersion trials for isothermal heavy gas releases. In addition, these authors have also dealt with nonisothermal dense gas releases [19]. The reliability of CFD simulation of cryogenic gas release was examined by Sklavounos and Rigas against field-scale experimental

FIGURE 2.2
Space Shuttle *Challenger* at lift-off, its smoke plume after in-flight breakup, and icicles that draped the Kennedy Space Center on that day. (From Great Images in NASA.)

FIGURE 2.7
Scene of a hydrogen tube trailer accident (From "H2Incidents Database" intended for public use, http://www.h2incidents.org/docs/265_1.pdf)

FIGURE 4.14
A photo from video footage taken March 14, 2011, shows the hydrogen explosion at the Fukushima Daiichi nuclear power complex at the moment of the event and afterward. (Reuters/NTV photo via Reuters TV. With permission.)

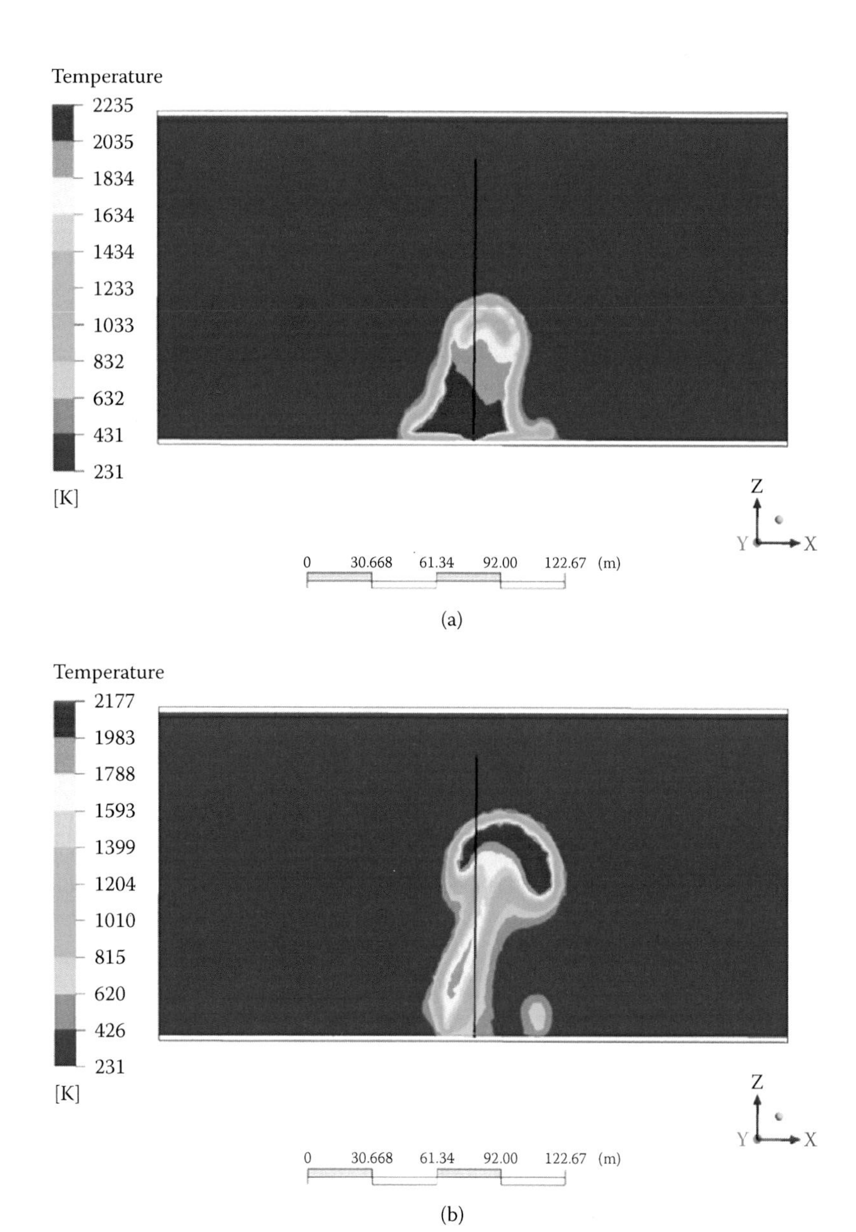

FIGURE 5.3
Fireball development after release and ignition of 1708 kg of propane. The burning cloud moves upward, obtaining positive buoyancy due to high temperatures, whereas air inflow from the left (inlet boundary condition) causes horizontal fireball shift to the right. (a) 1.2 s after ignition, (b) 3.2 s after ignition. (From Sklavounos, S., and Rigas, F., *Process Safety and Environmental Protection*. in press. doi:10.1016/j.psep.2011.06.008. With permission from Elsevier.)

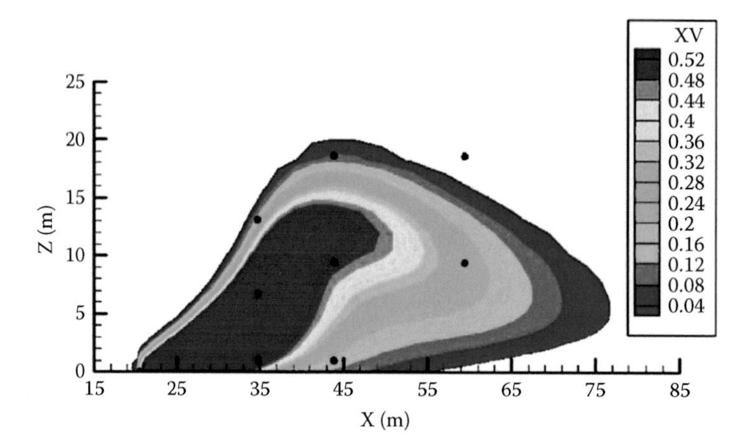

FIGURE 5.10
Predicted contours of hydrogen concentration (by volume) on symmetry plane at t = 21 s (Case 5: pool, with fence and contact heat transfer). (From Venetsanos, A.G., and Bartzis, J.G., *International Journal of Hydrogen Energy*, 32, 2171, 2007. With permission from Elsevier.)

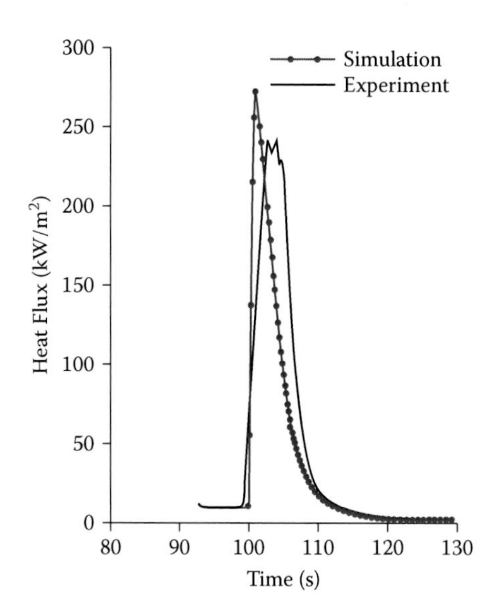

FIGURE 5.12
Experimental and simulated burning cloud snapshot for Coyote trial 3 (1 m height, 103 s). (From Rigas, F., and Sklavounos, S., *Chemical Engineering Science*, 61, 1444, 2006. With permission from Elsevier.)

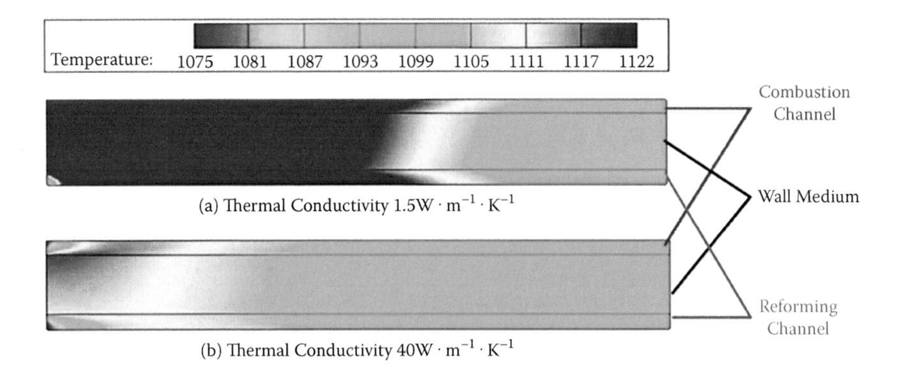

FIGURE 5.13
The temperature profiles inside the reactors with metallic and ceramic walls (residence time 10 ms, fuel gas: reforming gas [2:3, inlet temperature 1073 K, H_2O:CH_4 [3:1, absolute pressure 1 atm). (From Zhai, X. et al., *International Journal of Hydrogen Energy*, 35, 5383, 2010. With permission from Elsevier.)

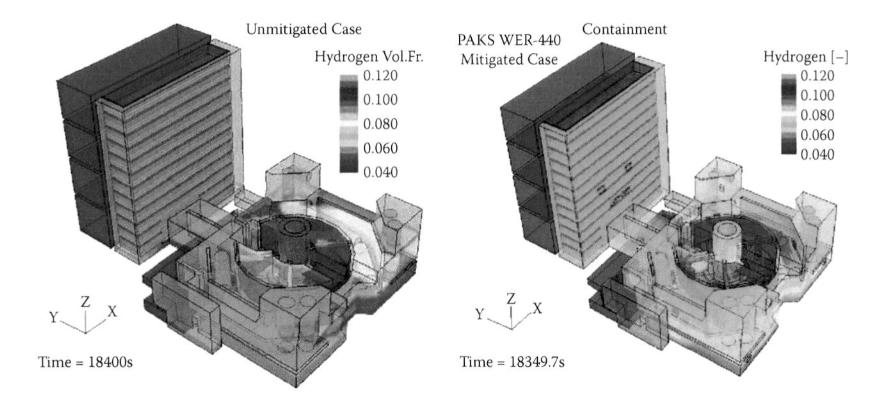

FIGURE 5.16
Hydrogen concentrations in the regions modeled at around 18,400 s. (From Heitsch, M., Huhtanen, R., Techy, Z., Fry, C. Kostka, P., CFD evaluation of hydrogen risk mitigation measures in a VVER-440/213 containment, Nuclear Engineering and Design, 240, 385, 2010) [44]. (With permission from Elsevier)

FIGURE 6.6
Hydrogen fireball about 170 m/sec after tank rupture in Test 2 (cylinder installed under an SUV). (From Zalosh, R., Proceedings of the 5th International Seminar on Fire and Explosion Hazards, Edinburgh, UK, April 23–27, 2007. With permission from the School of Engineering, University of Edinburgh, http://www.see.ed.ac.uk/feh5/pdfs/FEH_pdf_pp149.pdf.)

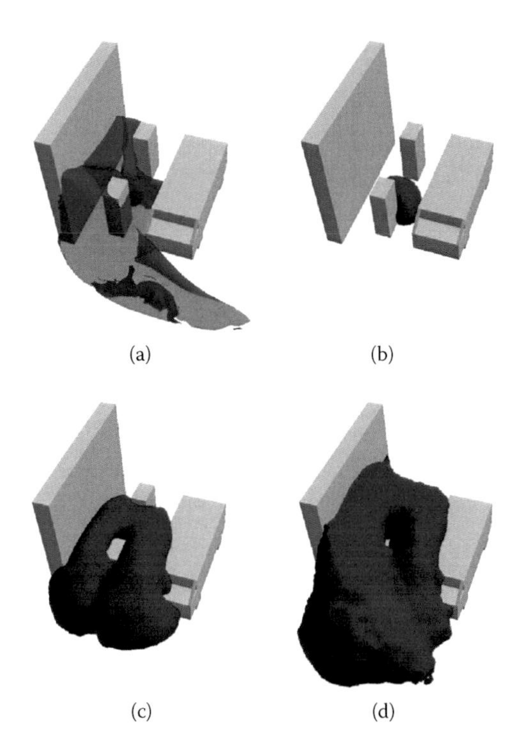

FIGVURE 6.9
Non-mitigated west wind scenario. Ignition is located between the dispensers. The molar fraction (0.04) in blue colour (a) shows the initial hydrogen flammable cloud. Isosurface of temperature (1000 K) in red colour (b,c,d) shows the propagation of the flame [25] (With permission from Elsevier)

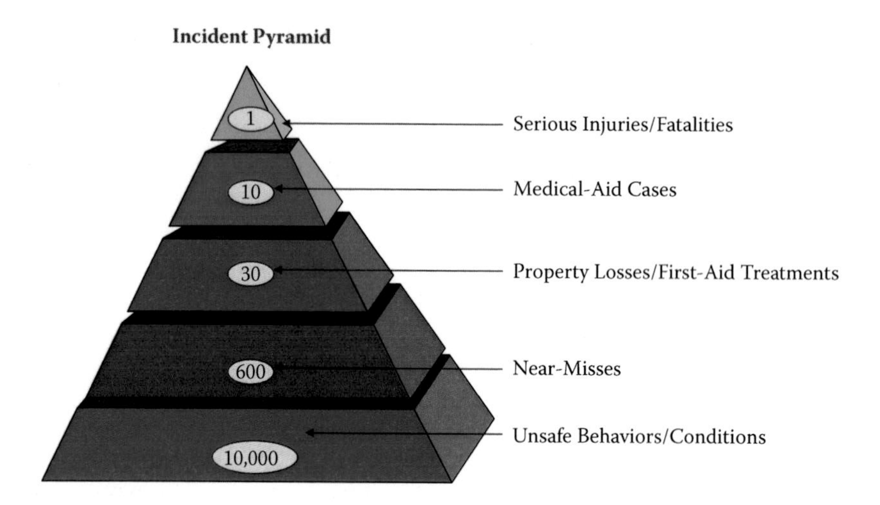

FIGURE 8.1
The incident pyramid. (From Creedy, G., *Process Safety Management*, PowerPoint presentation prepared for Process Safety Management Division, Chemical Institute of Canada, Ottawa, Ontario, 2004. With permission.)

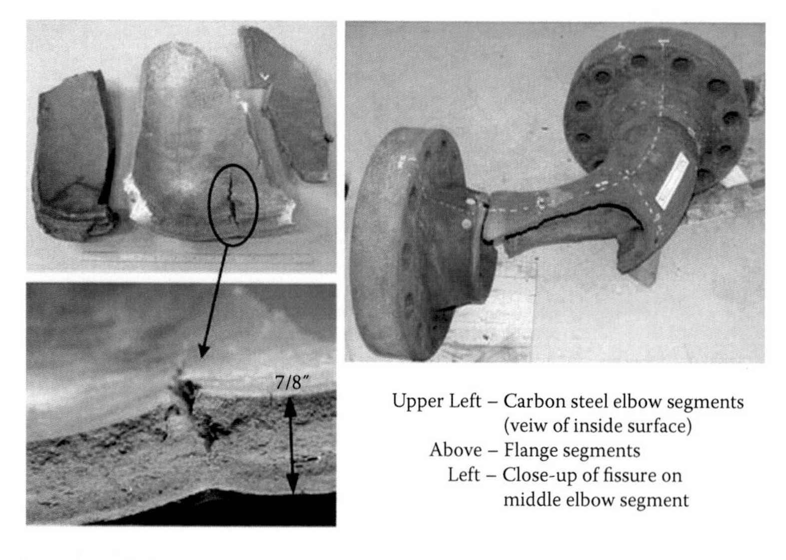

FIGURE 11.3
Ruptured carbon-steel pipe elbow subjected to high-temperature hydrogen attack (from CSB [8])

data including hydrogen and natural gas releases. A characteristic example of natural gas experiments is the Burro Series trials of LNG spills. The project was sponsored by the United States Department of Energy (U.S. DOE) and was conducted by LLNL and the Naval Weapons Center (NWC) at the China Lake site in California during the summer of 1980 to determine the transport and dispersion of vapor from spills of LNG on water. A detailed description of the experiments and analysis of the results are given in a report by LLNL [20]. Hydrogen dispersion experiments with data available to the public are the LH_2 spills conducted in 1980 by Witcofski and Chirivella [21] at NASA Langley Research Center at the White Sands test facility. The objective of the program was to study the phenomena associated with the rapid release of large amounts of LH_2, with emphasis on the generation and dispersion of the flammable hydrogen cloud.

Figures 5.5 and 5.6 show the experimental gas concentration histories versus the computational ones for the LNG Burro 8 trial and the LH_2 trial 6 obtained by Sklavounos and Rigas [19]. The experimental observations of the concentration change are found statistically to be in good agreement with the computationally predicted histories, showing that the CFD predictions are reasonable.

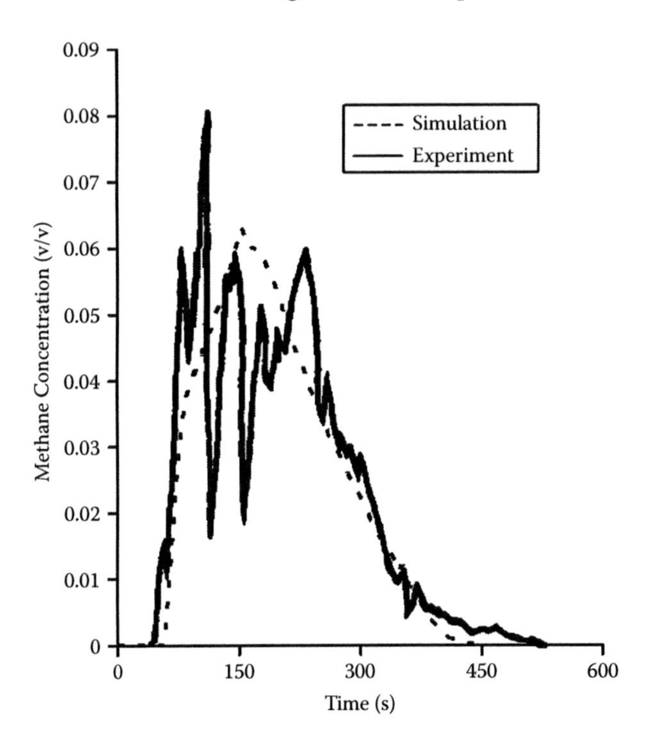

FIGURE 5.5
Computational and experimental methane concentration histories at point (140, 0, 1) for the Burro 8 trial. (From Sklavounos, S., and Rigas, F., *Energy and Fuels*, 19, 2535, 2005. With permission from Elsevier.)

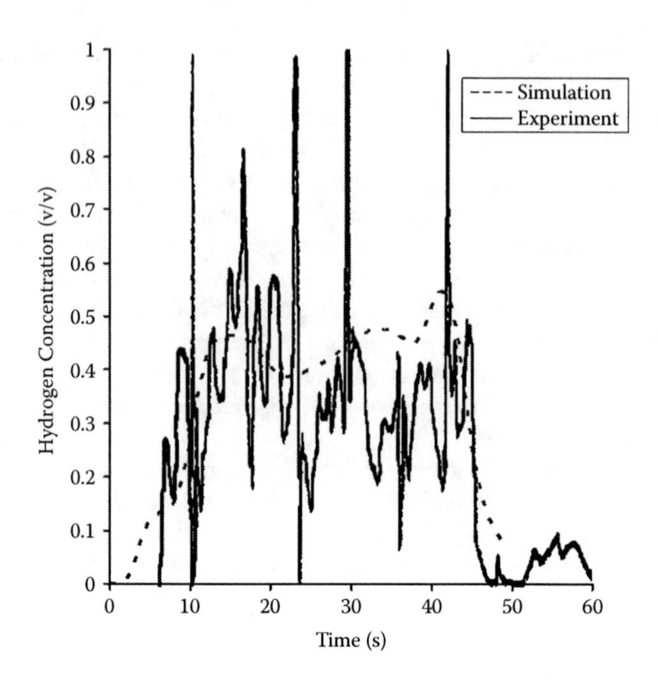

FIGURE 5.6
Computational and experimental hydrogen concentration histories at point (9.1, 0, 1) for the LH2 trial 6. (From Sklavounos, S., and Rigas, F., *Energy and Fuels*, 19, 2535, 2005. With permission from Elsevier.)

Further elaboration of simulation results and the calculation of specific statistical performance measures led to the conclusion that the predicted values were lying within a factor of two of the observed values of cloud passage duration, maximal concentration, and total dose; therefore, it was inferred that the predictions of CFX code were consistent with the data from the field tests.

The behavior of hydrogen during its atmospheric dispersion has also been examined by Rigas and Sklavounos [7] considering two different cases: isothermal release under pressurization conditions and cryogenic (nonisothermal) release under liquefaction conditions. Three-dimensional simulations were run for a typical accidental release scenario, which included hydrogen discharge under pressurization and liquefaction conditions in an imaginary computational domain with dimensions 160 m × 40 m × 30 m. Hydrogen release was modeled as a ground-level area source with inflow rate equal to 2 kg/s for 5 s, with typical atmospheric conditions (1 atm, 20°C), ground temperature equal to 15°C, and wind speed equal to 3 m/s.

The results demonstrated substantial differentiation between pressurized and cryogenic hydrogen release as far as cloud dispersion is concerned. In the former case, the cloud lifts immediately, moving almost vertically (Figure 5.7a through 5.7d), whereas in the latter, it travels downwind, almost horizontally remaining close to the ground (Figure 5.8a through 5.8d).

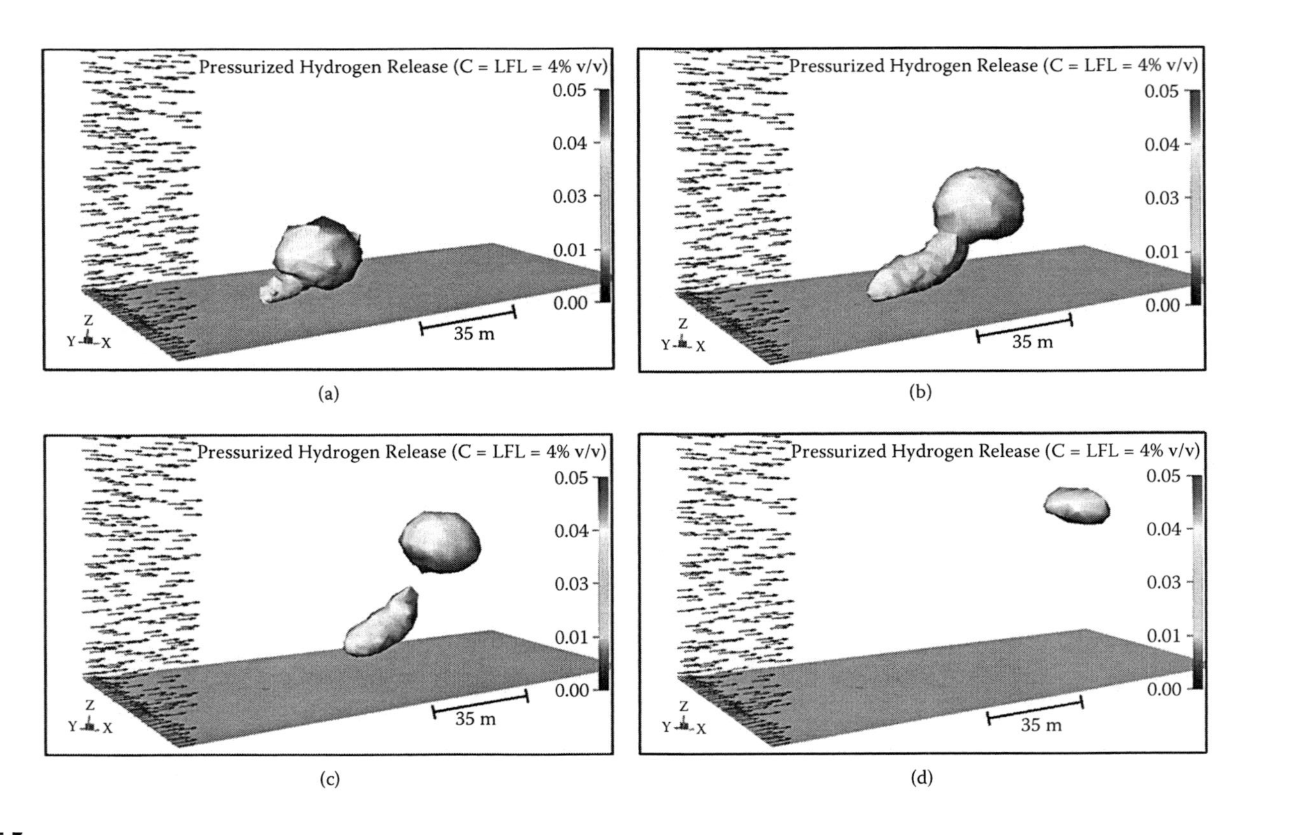

FIGURE 5.7
Simulated dispersion snapshots following pressurized hydrogen release at times (a) 2.4 s, (b) 6.2 s, (c) 9.6 s, and (d) 12.4 s after start of release. (From Rigas, F., and Sklavounos, S., *International Journal of Hydrogen Energy*, 30, 1501, 2005. With permission from Elsevier.)

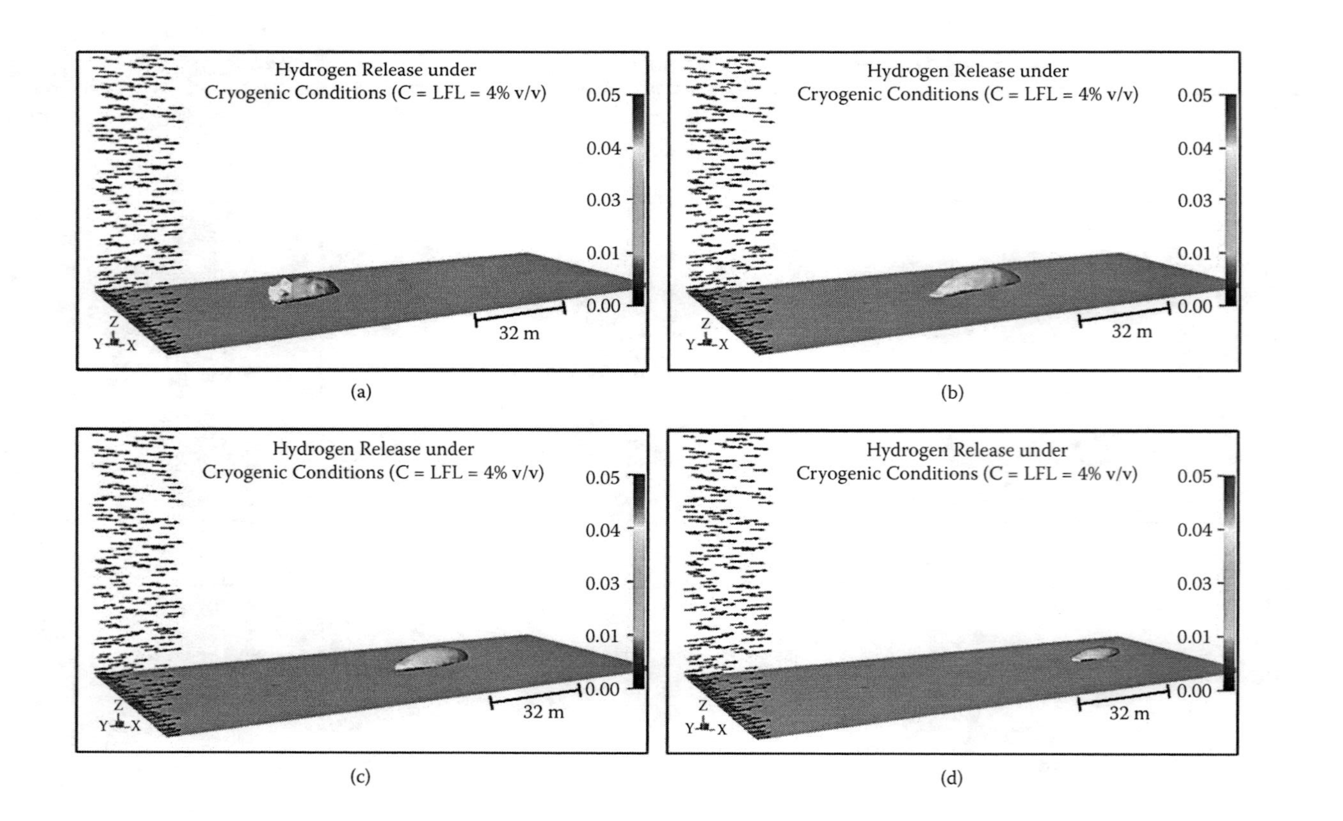

FIGURE 5.8
Simulated dispersion snapshots following cryogenic hydrogen release at times (a) 3.5 s, (b) 10 s, (c) 13 s, and (d) 18.5 s after start of release. (From Rigas, F., and Sklavounos, S., *International Journal of Hydrogen Energy*, 30, 1501, 2005. With permission from Elsevier.)

It can be seen from Figures 5.9a and 5.9b that spreading of a cryogenic-origin hydrogen cloud simulated by Rigas and Sklavounos [7] is similar to that of liquefied natural gas (LNG) that has been experimentally observed in field-scale trials at LLNL installations [16], namely horizontal cloud shift at low height. This is the result of the greater inertia of the heavy cryogenic gas that reduces the rate of turbulent mixing, thus leading to delayed dissolution

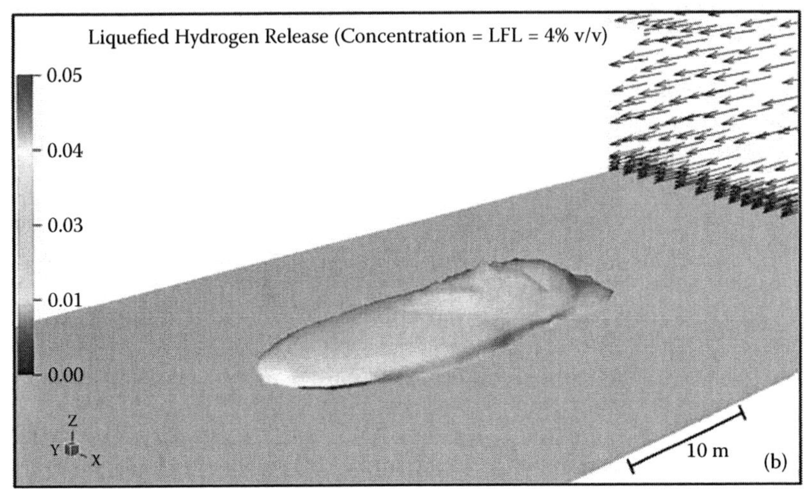

FIGURE 5.9
Qualitative comparison of dispersing cloud following (a) liquefied natural gas release; experimental record from Burro series trials; and (b) liquefied hydrogen release; computational estimation 7 s after release initiation. (a) From Lees, F.P., *Loss Prevention in the Process Industries*, Butterworth Heinemann, Oxford, 1996. With permission. (b) From Rigas, F., and Sklavounos, S., *International Journal of Hydrogen Energy*, 30, 1501, 2005. With permission from Elsevier.)

times. In addition, the greater density of the cloud produces a gravity-driven flow that tends to reduce the cloud height and increase its width. Thus, in contrast to lighter-than-air gas, the persistence of heavy cloud at low height significantly increases the risk of ignition.

The simulation of atmospheric dispersion of clouds from liquefied *cryogenic hydrogen releases* by Rigas and Sklavounos [7] showed that hydrogen disperses as a heavier-than-air gas when discharged in cryogenic conditions, in contrast with a high-pressure gas release that produces a lighter-than-air cloud. In the former case, hydrogen remains within the LFL range close to the ground for some time, increasing the risk for various types of fires and explosions. Visualization of computer-simulated cryogenic hydrogen dispersions demonstrated that hydrogen cloud behavior in this case is similar to that observed during LNG experimental releases. Consequently, accidental liquefied hydrogen release scenarios should be addressed as heavier-than-air releases using the appropriate dispersion models for calculating heavy and not light gas dispersion. Moreover, this change of the behavior of a light gas on dispersion during cryogenic release should be kept in mind in hazard evaluation procedures and when establishing safety measures.

In practical applications, dispersion calculations need to be performed in *topographically complex environments*. Thus, solid obstacles intervening in the dispersion area should be considered in the computations. In previous validation work, it has been proven that CFD codes constitute powerful tools for complex terrain dispersion simulation providing very accurate results with excellent visualization capabilities, which can be helpful in quantitative risk analysis applications [18]. The dominating mixing mechanism between gas and air, which is the turbulent flow developing in the vicinity of the solid obstacle, was then computationally approached satisfactorily by several turbulence models: the k-ε and shear stress transport (SST) models showed improved robustness during solver processing. The SSG model entailed increased central processing unit (CPU) time without significant enhancement of accuracy of results. The SSG, k-ε, and SST models appeared to overestimate maximum concentrations recorded in the trials, whereas the k-ω model underestimated them.

Generally, the results obtained through the numerical simulations showed good agreement compared with the experimental data, leading to the conclusion that CFD techniques can be effectively used in consequence assessment procedures concerning toxic/flammable dispersion scenarios in real terrains, where box-models have limited capabilities.

Venetsanos and Bartzis [22] have also investigated CFD modeling of large-scale LH_2 spills in an unobstructed environment using the ADREA-HF code. The authors performed simulations to investigate the effects of the source model (jet or pool), the modeling of the earthen sides of the pond around the source, and the ground heat transfer. Comparison of the predicted hydrogen concentrations against the experimental results from NASA's trial-6 experiment at the available sensor locations was done. Generally, simulation jet

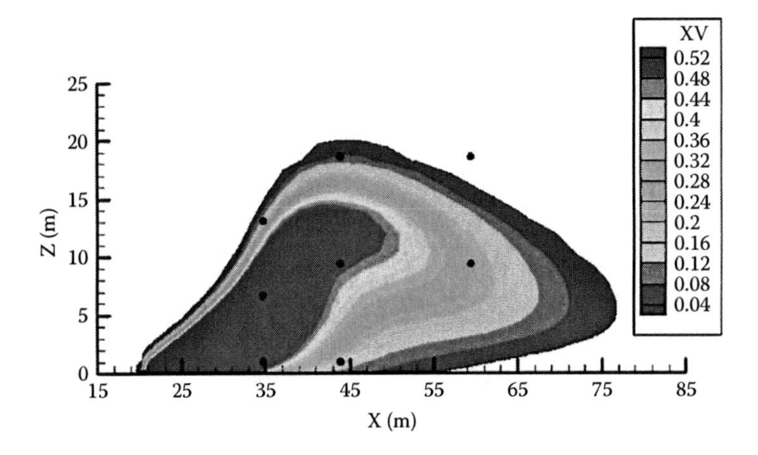

FIGURE 5.10 (See color insert.)
Predicted contours of hydrogen concentration (by volume) on symmetry plane at t = 21 s (Case 5: pool, with fence and contact heat transfer). (From Venetsanos, A.G., and Bartzis, J.G., *International Journal of Hydrogen Energy*, 32, 2171, 2007. With permission from Elsevier.)

modeling was not found to be in good agreement with experimental data, with the exception of the simulation based on modeling the source as a two-phase jet pointing downwards, which resulted in predicted concentrations in much better agreement with the experiments. Modeling the source as a pool resulted in overestimation of concentration levels. An illustration of the predicted contours of hydrogen concentration (by volume) on a symmetry plane at 21 s from a hydrogen pool is shown in Figure 5.10.

The DREA-HF CFD code has also been investigated as a prediction tool for consequence assessment of hydrogen applications by Venetsanos, Papanikolaou, and Bartzis [23]. To this end, various types of hydrogen release scenarios were considered, including gaseous and liquefied releases, open, semiconfined, and confined environments, and sonic (under-expanded) and low momentum releases.

5.5.3 Releases in Confined Spaces

Swain et al. [24] have proposed the hydrogen risk assessment method (HRAM) to be used for establishing venting in buildings that contain hydrogen-fueled equipment or for determining optimum hydrogen sensor locations. The method can be used to reduce CFD modeling and utilize helium as a hydrogen surrogate to limit hazardous validation tests for hydrogen leakage into simple geometric enclosures.

An intercomparison exercise on the capabilities of CFD models to predict the short- and long-term distribution and mixing of hydrogen in confined spaces, as in a garage, was performed by Venetsanos et al. [25]. The rectangular volume studied was 78.4 m^3. The project engaged 12 different organizations, 10 different CFD codes, and 8 different turbulence models. The authors found large variations in predicted results in the first phase of the benchmark among the various modeling approaches and this was attributed to differences in turbulence models and numerical accuracy options (time/space resolution and discretization schemes). Nevertheless, during the second phase of the benchmark the variation between simulated and predicted results was reduced. (See also Chapter 9.)

Papanikolaou et al. [26] have studied numerically the release of hydrogen inside a naturally ventilated residential garage with an internal volume of 66.3 m^3. For validation purposes they used past experiments performed by Swain et al. [24] with helium as a surrogate gas. This was, in fact, an inter-laboratory study checking results from four different CFD packages, namely, ADREA-HF, FLACS, FLUENT, and CFX, with the experimental measurements inside the garage. Agreement between the partners' predictions and the experimental data was generally good. Yet, it was determined there was a tendency to overestimate the results of the upper sensors for the small and medium vent sizes and underestimate for the large vent size. The results of the lower sensors were generally over-predicted for the small and medium vent sizes whereas for the larger vent size the four codes either over-predicted or under-predicted at the sensor close to the leak and over-predicted at the sensor close to the vent.

In another article on hydrogen release and dispersion in enclosed spaces, Zhang et al. [27] compared experimental data with CFD computed results and the calculations of an analytical method. They found good agreement of all three approaches, provided a suitable value is given to Smagorinsly's constant to obtain "fine tuning" of the CFD application. The investigated space was a rectangular room with a volume of 78.4 m^3. The Fire Dynamics Simulator (FDS) code (a large-eddy simulation based code) available from NIST (National Institute of Standards and Technology) was used for CFD simulations [28].

Papanikolaou et al. [29] have obtained CFD simulations on small hydrogen releases inside a ventilated facility providing an internal volume of 25 m^3 and, thus, assessing ventilation efficiency. The study focused on the safety assessment of a facility hypothetically hosting a small H_2 fuel cell system. Dispersion experiments to validate the simulations were also performed. Simulation was accomplished by using the ADREA-HF CFD code. In general, good agreement was found between predicted and experimentally measured concentration/time histories for all simulated cases.

Venetsanos et al. [30] summarized the results from the project InsHyde on the use of hydrogen in confined spaces. (Again, see also Chapter 9.) The project investigated realistic small to medium indoor hydrogen leaks and provided recommendations for the safe use and storage of indoor hydrogen systems using experimental and simulation approaches. The authors used hydrogen and nitrogen for their experiments to evaluate short- and long-term dispersion patterns in garage-like settings. In addition, they performed combustion experiments to evaluate the maximum amount of hydrogen that could be safely ignited indoors and they extended their study to the ignition of larger amounts of hydrogen in obstructed environments outdoors. Their work included pretest simulations, validation of the available CFD codes against previously performed experiments with significant CFD code intercomparisons, as well as CFD application to investigate specific realistic scenarios.

5.5.4 Thermal Hazards from Hydrogen Clouds

Atmospheric releases of flammable gases such as hydrogen may lead to major fires with extensive effects on the surroundings, mainly due to the intense thermal load emissions. In activities where hazards are associated with cloud fires, there is the need for societal risk assessment that involves the estimation of hazardous zones due to the resulting thermal radiation. However, until now limited work has been done on modeling the effects of flash fires; available techniques have been judged insufficient by the Center for Chemical Process Safety, CCPS [6].

A methodology for modeling *flash fire radiation* was developed by Raj and Emmons and proposed by CCPS [6], in which a flash fire is modeled as a two-dimensional, turbulent flame propagating at constant speed. Yet, the model requires simplifying assumptions (i.e., stationary location of burning cloud) and presents certain deficiencies (i.e., ignoring of ignition site location). Moreover, a multipoint grid-based approach for radiation impact analysis was adopted by Pula et al. [31] providing estimations of radiation effects at different locations in the area of interest. A three-dimensional computational approach based on fluid dynamics techniques was recently attempted, aiming at the estimation of resulting thermal radiation emissions and overpressure in large-scale cloud fires [32, 33].

The experimentation in the field of gas cloud fires appears to be limited. A unique set of large-scale experiments that involve the release, dispersion, ignition, and combustion of flammable natural gas clouds in the open air does, however, exist, with the code name Coyote. Coyote series trials conducted by LLNL in 1983 at the Nevada test site provided an integrated dataset for use in validation studies [34, 35]. The objective of the experiments was to determine the transport and dispersion of vapors from LNG spills, and in addition to investigate the damage potential of vapor cloud fires. Transient simulations of the flammable cloud combustion accomplished by Rigas and Sklavounos [32, 33] and performed with the CFX (CFD code referred to above) provided reasonable estimations of the thermal flux histories (Figure 5.11), whereas the predictions of the thermal impulses proved to be statistically valid and consistent with the experimental results obtained at LLNL installations [34], as verified by strict statistical tests.

In addition, post processing of the computational results resulted in visualizations of the combustion progress within the cloud. Thus, comparisons between the computational and the experimentally observed combustion

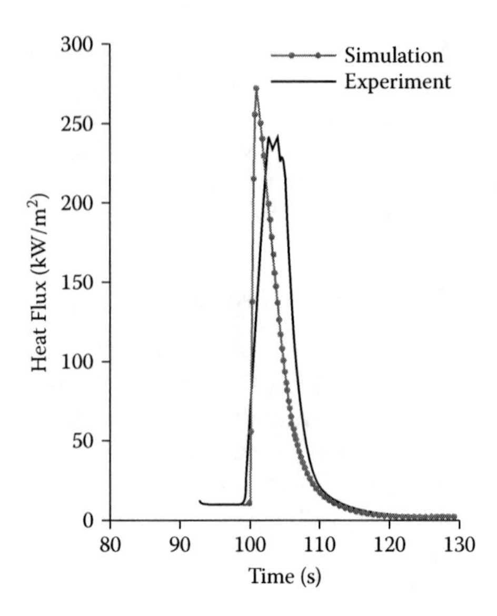

FIGURE 5.11
Experimental versus computational heat flux histories for Coyote trial 3 at position (65, 57, 1). (From Rigas, F., and Sklavounos, S., *Chemical Engineering Science*, 61, 1444, 2006. With permission from Elsevier.)

regions were obtained, revealing good agreement between the corresponding snapshots (Figure 5.12). Generally, it was concluded that the evolution of a gas cloud fire and the oncoming consequences may be efficiently simulated utilizing CFD techniques.

5.5.5 Simulations in the Chemical Industries

Chemical industries (including the petroleum industry) produce and consume huge quantities of hydrogen for their needs, mainly for upgrading fossil fuels and for the production of ammonia. As a result, many accidents have occurred in these industries due to the use of hydrogen. To counteract these occurrences, many publications have recently stressed the use of CFD tools to simulate complicated chemical and physical phenomena with the aim of preventing loss of control of chemical reactors.

Zhai et al. [36] have proposed CFD simulation with the detailed chemistry of steam reforming of methane (SRM) for hydrogen production in an integrated microreactor. The authors claim that steam reforming of methane in microreactors has great potential to develop a low-cost, compact process for hydrogen production via a shortening of reaction time from seconds to milliseconds. They simulated a microreactor design for SRM reaction with the integration of a microchannel for Rh-catalyzed endothermic reactions, a microchannel for Pt-catalyzed exothermic reactions, and a wall in between with a Rh- or Pt-catalyst–coated layer. They accomplished CFD modeling

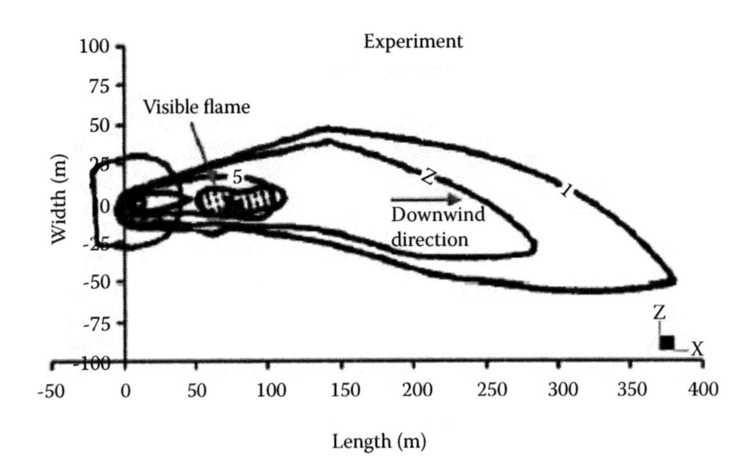

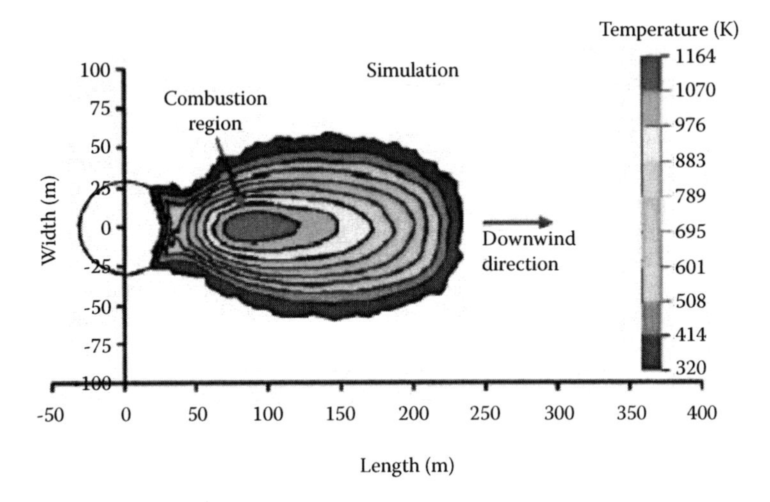

FIGURE 5.12 (See color insert.)
Experimental and simulated burning cloud snapshot for Coyote trial 3 (1 m height, 103 s). (From Rigas, F., and Sklavounos, S., *Chemical Engineering Science*, 61, 1444, 2006. With permission from Elsevier.)

by adopting elementary reaction kinetics for the SRM process in the CFD model and describing combustion in the channel by global reaction kinetics. The model predictions were validated against experimental data from the literature. For the extremely fast reactions in both channels, the simulations indicated the significance of the heat conduction ability of the reactor wall, as well as the interplay between the exothermic and endothermic reactions (e.g., the flow rate ratio of fuel gas to reforming gas). This approach provides a potential way of efficiently investigating and controlling these highly exothermic and hazardous reactions, aiming at developing more economical hydrogen production.

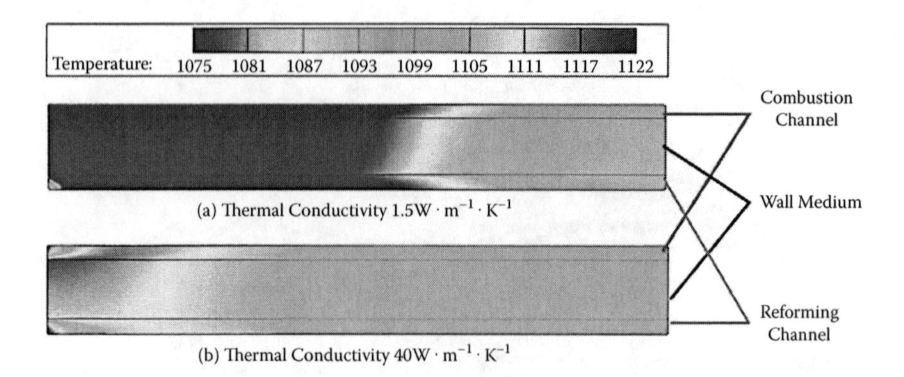

FIGURE 5.13 (See color insert.)
The temperature profiles inside the reactors with metallic and ceramic walls (residence time 10 ms, fuel gas: reforming gas [2:3, inlet temperature 1073 K, $H_2O:CH_4$ [3:1, absolute pressure 1 atm). (From Zhai, X. et al., *International Journal of Hydrogen Energy*, 35, 5383, 2010. With permission from Elsevier.)

The reactor performance was investigated using a metallic wall with thermal conductivity of 40 W m^{-1} K^{-1} and a ceramic wall with thermal conductivity of 1.5 W m^{-1} K^{-1} as shown in Figure 5.13.

Successful CFD simulation has also been accomplished by Dou and Song [37] on hydrogen production from steam reforming of glycerol in a fluidized bed reactor. This conversion is of great environmental value considering the high surplus of glycerol now produced from the conversion of vegetable oils or animal fats to biodiesel fuel. Thus, glycerol is becoming a cheap market product and is expected to become a waste problem. The FLUENT code was used in these simulations with the following three-step reaction scheme:

$$C_3H_8O_3 + H_2O \leftrightarrow CH_4 + 2CO_2 + 3H_2$$

$$CH_4 + H_2O \leftrightarrow CO + 3H_2$$

$$CO + H_2O \leftrightarrow CO_2 + H_2$$

Chemical thermodynamics and experimentation revealed that to avoid coke deposition on the Ni-based catalyst, the reaction temperature should be raised to 600°C. Thus, this CFD simulation is a valuable tool to predict and prevent hazardous conditions when applying this new chemical process industrially.

To counteract hydrogen deflagrations in the process industries, Baraldi et al. [38] utilized the results of the experiments performed by Pasman and Groothuisen [39] on explosions of stoichiometric hydrogen-air mixtures in a cylinder-type combustion vessel of a volume of 0.95 m^3 equipped with pressure relief vents. The authors simulated experimental results with a discrepancy below 20 percent for the case of a 0.2 m^2 vent, whereas the discrepancy

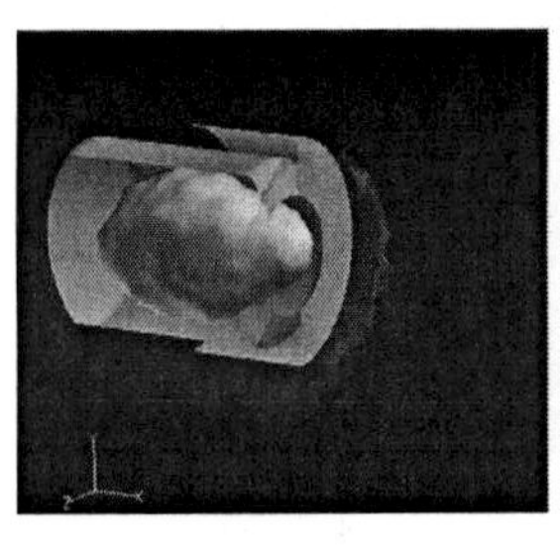
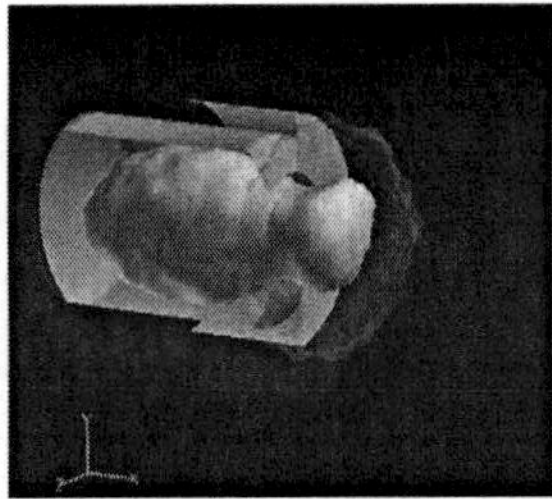
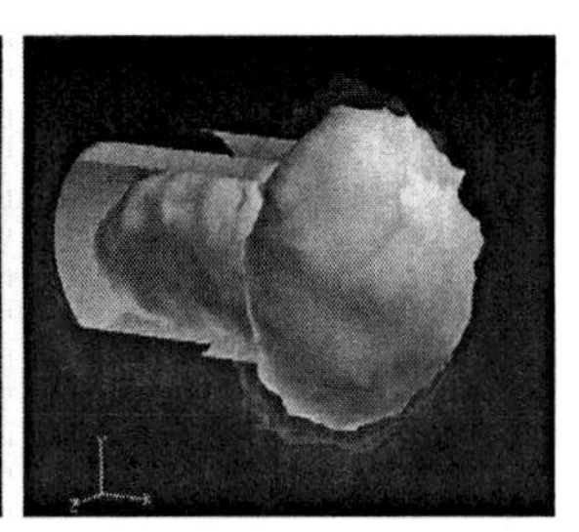

FIGURE 5.14
Flame propagation of a stoichiometric mixture of hydrogen-air through a 0.3 m^2 vent. (From Baraldi, D. et al., *International Journal of Hydrogen Energy*, 35, 12381, 2010. With permission from Elsevier.)

with a 0.3 m^2 vent was much higher (up to 45 percent). The flame propagation for the case with a 0.3 m^2 vent is illustrated in Figure 5.14.

5.5.6 Simulations in the Nuclear Industry

A severe hazard in nuclear facilities, in the event of loss of control, is the accidental production of large quantities of hydrogen via various mechanisms. Fast reactions, for example, in a few minutes, that may contribute to this phenomenon include zirconium-steam reactions, steel-steam reactions, molten core-water reactions, and molten core-concrete reactions. On the other hand, slow reactions may produce considerable quantities of hydrogen. Such reactions include water radiolysis, and corrosion of aluminum, zinc-based paint, and galvanized steel. Mixing of hydrogen with air in a confined space will result in flammable or even detonable mixtures taking into account the wide flammability and detonability limits of hydrogen-air mixtures. This eventuality has occurred in many cases of nuclear accidents, including the catastrophic hydrogen explosion at the Fukushima Daiichi nuclear power complex in Japan on March 14, 2011.

In an effort to prevent such accidents, Baraldi, Heitsch, and Wilkening [40] performed CFD simulations of hydrogen deflagrations in a simplified EPR (European pressure reactor) containment, aiming at constructing prediction models for the overpressures and high temperatures generated from these events. In this multiparametric analysis, the authors investigated pressure differences on the two sides of walls, as well as various geometrical configurations considering the number of vents, size, and position. Single and multiple ignition points were also investigated. CFX and REACFLOW codes were used in this work and useful results were obtained.

Redlinger [41] validated DET3D (a CFD tool) against experimental and theoretically calculated results aiming at investigating and improving nuclear reactor safety. The author simulated hydrogen detonations in complex three-dimensional geometries occurring in assumed accident scenarios in nuclear reactors. The potential of hydrogen-air mixtures for flammability, flame acceleration, deflagration-to-detonation transition, and detonability was checked using the σ (expansion ratio) and λ (detonation cell size) criteria of mixtures, as described by Dorofeev et al. [42]. Then the detonability conditions were used as initial data to run the DET3D code. Good agreement was found between computations and experiments in all studied cases.

The GASFLOW CFD code was applied by Xiong, Yang, and Cheng [43] in an effort to investigate hydrogen risk at the Qinshan-II nuclear power plant in China. The authors analyzed the effect of water spray on hydrogen risk in the containment considering a large break loss of coolant accident (LBLOCA). They selected three different spray strategies, namely, without spray, with direct spray, and with both direct and recirculation spray. They determined a significant influence of spray modes on hydrogen distribution inside the containment. The efficiency of the passive autocatalytic recombiners (PAR) was also investigated. No substantial influence by spray modes was observed. The results obtained by a CFD analysis of a new PAR model were in good agreement with the experimental results.

Heitsch et al. [44] performed simulations of the Paks Nuclear Power Plant using GASFLOW, FLUENT, and CFX codes by modeling a defined severe hydrogen release accident scenario. The authors investigated the risk of losing integrity of containment after hydrogen deflagrations. The mitigation measure proposed and assessed was the implementation of catalytic recombiners in the plant. Both unmitigated and recombiner-mitigated simulations were performed and the results were compared. Simulations of full containment subject to a severe accident were not possible until recently due to the prohibitive computing times and the limited capabilities of the codes; yet this work proved that this is now possible with improved codes and more powerful computers. Just the same, the wall clock time needed for each of the simulations in this work was around 40 days, even when six to eight processors were used in parallel. The analysis showed that at least 30 recombiners are needed if it is desirable to avoid significant flammable clouds with more than 8 percent of hydrogen being formed. The main compartments of the containment modeled in the CFD simulations are shown in Figure 5.15. Figure 5.16 illustrates the reduction of hydrogen in the containment for both unmitigated and recombiner-mitigated cases.

Prabhudharwadkar and Iyer [45] formulated a model compatible with a CFD code to simulate hydrogen distribution using a passive catalytic recombiner in a nuclear power plant containment aiming at preventing hydrogen-air explosions. This was an effort to counteract the difficulties in modeling resulting from the much smaller size of a catalytic recombiner (needing very fine mesh size inside the recombiner channels) compared to the containment. After conducting a parametric study, the authors modeled the recombiner channels as single computational cells. This avoids full resolution of these channels, allowing the use of a

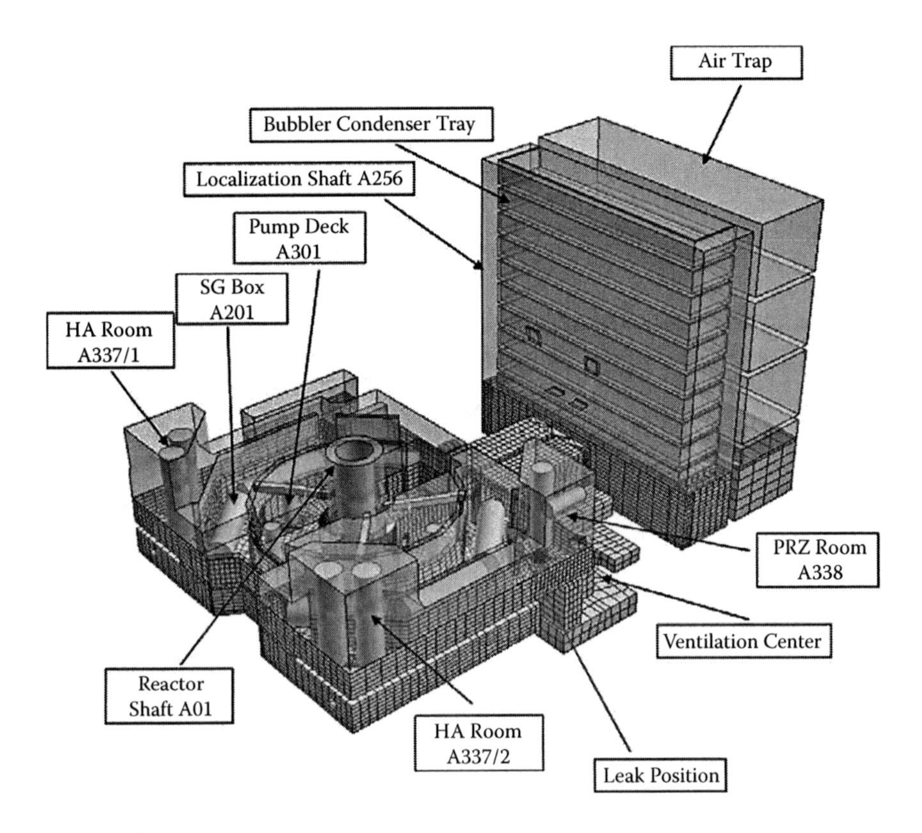

FIGURE 5.15
Regions modeled in the CFD simulations. (From Heitsch, M. et al., *Nuclear Engineering and Design*, 240, 385, 2010. With permission from Elsevier.)

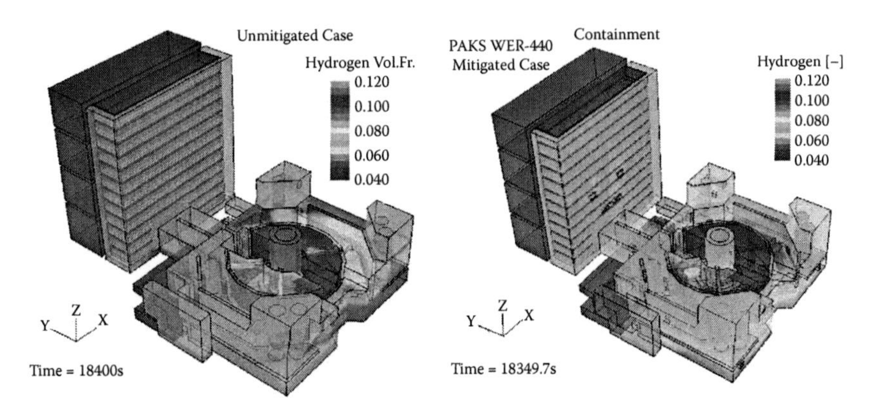

FIGURE 5.16 (See color insert.)
Hydrogen concentrations in the regions modeled at around 18,400 s. (From Heitsch, M., Huhtanen, R., Techy, Z., Fry, C., Kostka, P., CFD evaluation of hydrogen risk mitigation measures in a VVER-440/213 containment, *Nuclear Engineering and Design*, 240, 385, 2010) [44]. (With permission from Elsevier)

one-node recombiner channel modeling approach, thus offering a great gain in computational time. Predictions with the single node approach were within 5 percent accuracy for steady-state cases when compared to the detailed simulations.

Ferng and Chen [46] analyzed the thermal-hydraulic characteristics and hydrogen generation from graphite-water reactions after a steam generator tube rupture in a HTR-10 reactor (a graphite-moderated and helium-cooled reactor with a pebble-bed core). They developed a transient three-dimensional compressible CFD model to investigate four postulated accidents in HTR-10 considering ruptures in one or two steam generator tubes which cause hydrogen and carbon monoxide generation. The computations showed that the hydrogen quantity produced from the reaction of water with graphite is not sufficient to reach the flammability limits of hydrogen-air mixtures. In fact, the maximum quantity of hydrogen that may be produced from two tube ruptures is a volume fraction of about 0.5 percent, which is much less than the lower flammability limit of 4 percent. In addition, the maximum temperature of the nuclear fuel in the reactor core was computed to be 1218 K, much less than the limiting temperature (1873 K) for graphite. The authors therefore concluded that the nuclear fuel would not be damaged under the worst case conditions.

References

1. Walt Pyle, Hydrogen Storage, *Home Power*, 59, 14–15, June/July, 1997.
2. WolframAlpha, Computational Knowledge Engine (http://www.wolframalpha.com).
3. Verfondern, K., *Safety Considerations on Liquid Hydrogen*, Forschungszentrum Juelich GmbH, Energy and Environment, Vol. 10, 2008, pp. 25–27.
4. Tzimas, E., Filiou, C., Peteves, S.D., and Veyret, J.B., *Hydrogen Storage: State-of-the-Art and Future Perspective*, European Commission, Directorate General Joint Research Centre, Petten, The Netherlands, 2003.
5. Warren, P., *Hazardous Gases and Fumes*, Butterworth-Heinemann, Oxford, 1997, p. 96.
6. CCPS (Center for Chemical Process Safety), *Guidelines for Evaluating the Characteristics of Vapor Cloud Explosions, Flash Fires and BLEVEs*, American Institute of Chemical Engineers, New York, 1994, p. 158.
7. Rigas, F., and Sklavounos, S., Evaluation of hazards associated with hydrogen storage facilities, *International Journal of Hydrogen Energy*, 30, 1501, 2005.
8. D.R. Lide, Ed., *Handbook of Chemistry and Physics*, 75th ed., CRC Press, Boca Raton, FL, Chap. 6, 1994.
9. Shebeko, Yu. N., Shevchuck, A.P., and Smolin, I.M., BLEVE prevention using vent devices, *Journal of Hazardous Materials*, 50, 227, 1996.

10. Sklavounos, S., and Rigas, F., Advanced multi-perspective computer simulation as a tool for reliable consequence analysis, *Process Safety and Environmental Protection*, 90(2), 129, 2012.
11. Pohanish, R.P., and Green, S.A., *Rapid Guide to Chemical Incompatibilities*, Van Nostrand Reinhold, New York, 1997, p. 434.
12. Carson, P., and Mumford, C., *Hazardous Chemicals Handbook*, Butterworth-Heinemann, Oxford, 1994, pp. 181, 202.
13. Taylor, J.R., *Risk Analysis for Process Plant, Pipelines and Transport*, Chapman & Hall, London, 1994, p. 102.
14. CCPS, *Guidelines for Hazard Evaluation Procedures*, American Institute of Chemical Engineers, New York, 1992, p. 69.
15. HSE (Health and Safety Executive), *Application of QRA in Operational Safety Studies*, Health and Safety Executive report 025, 2002, p. 19.
16. LLNL (Lawrence Livermore National Laboratory), *Burro Series Data Report*, LLNL/NWC report UCID-19075 Vol. 1, LLN, Berkeley, CA, 1982.
17. Versteeg, H.K., and Malalasekera, W., *An Introduction to Computational Fluid Dynamics: The Finite Volume Method*, Longman, New York, 1995, p. 85.
18. Sklavounos, S., and Rigas, F., Validation of turbulence models in heavy gas dispersion over obstacles, *Journal of Hazardous Materials*, 108, 9, 2004.
19. Sklavounos, S., and Rigas, F., Fuel gas dispersion under cryogenic release conditions, *Energy and Fuels*, 19, 2535, 2005.
20. LLNL, *Description and Analysis of Burro Series 40-m3 LNG Spill Experiments*, LLNL/NWC report No. UCRL-53186, Lawrence Livermore National Laboratory, Berkeley, CA, 1981.
21. Witcofski, R.D., and Chirivella, J.E., Experimental and analytical analyses of the mechanisms governing the dispersion of flammable clouds formed by liquid hydrogen spills, *International Journal of Hydrogen Energy*, 9, 425, 1984.
22. Venetsanos, A.G., and Bartzis, J.G., CFD modeling of large-scale LH2 spills in open environment, *International Journal of Hydrogen Energy*, 32, 2171, 2007.
23. Venetsanos, A.G., Papanikolaou, E., and Bartzis, J.G., The ADREA-HF CFD code for consequence assessment of hydrogen applications, *International Journal of Hydrogen Energy*, 35, 3908, 2010.
24. Swain, M.R. et al., Hydrogen leakage into simple geometric enclosures, *International Journal of Hydrogen Energy*, 28, 229, 2003.
25. Venetsanos, A.G., Papanikolaou E., Delichatsios, M., Garcia, J., Hansen, O.R., Heitsch, M., Huser, A., Jahn, W., Jordan, T., Lacome, J.M., Ledin, H.S., Makarov, D., Middha P., Studer, E., Tchouvelev, A.V., Teodorczyk, A., Verbecke, F., and Van der Voort, M.M., An inter-comparison exercise on the capabilities of CFD models to predict the short and long term distribution and mixing of hydrogen in a garage, *International Journal of Hydrogen Energy*, 34, 5912, 2009.
26. Papanikolaou, E., Venetsanos, A.G., Heitsch, M., Baraldi, D., Huser, A., Pujol, J., Garcia, J., and Markatos, N., HySafe SBEP-V20: Numerical studies of release experiments inside a naturally ventilated residential garage, *International Journal of Hydrogen Energy*, 35, 4747, 2010.
27. Zhang, J., Delichatsios, M.A., and Venetsanos, A.G., Numerical studies of dispersion and flammable volume of hydrogen enclosures, *International Journal of Hydrogen Energy*, 35, 6431, 2010.

28. FDS—Fire Dynamics Simulator, available from NIST (National Institute of Standards and Technology) at: http://fire.nist.gov/fds.
29. Papanikolaou, E., Venetsanos, A.G., Cerchiara, G.M., Carcassi, M., and Markatos, N., CFD simulations on small hydrogen releases inside a ventilated facility and assessment of ventilation efficiency, *International Journal of Hydrogen Energy*, 36, 2597, 2011.
30. Venetsanos, A.G., Adams, P., Azkarate, I., Bengaouer, A., Brett, L., Carcassi, M.N., Engebø, A., Gallego, E., Gavrikov, A.I., Hansen, O.R., Hawksworth, S., Jordan, T., Kesslerm, A., Kumar, S., Molkov, V., Nilsen, S., Reinecke, E., Stoecklin, M., Schmidtchen, U., Teodorczyk, A., Tigreat, D., and Versloot, N.H.A., On the use of hydrogen in confined spaces: Results from the internal project InsHyde, *International Journal of Hydrogen Energy*, 36, 2693, 2011.
31. Pula,. R., Khan, F., Veitsch, B., and Amyotte, P., Revised fire consequence models for off-shore quantitative risk assessment, *Journal of Loss Prevention in the Process Industries*, 18, 443, 2005.
32. Sklavounos, S., and Rigas, F., Simulation of Coyote series trials, Part I: CFD estimation of non-isothermal LNG releases and comparison with box-model predictions, *Chem. Eng.Sci.*, 61, 1434, 2006.
33. Rigas, F., and Sklavounos, S., Simulation of Coyote series trials – Part II: a computational approach to ignition and combustion of flammable vapor clouds, *Chemical Engineering Science*, 61, 1434, 2006.
34. LLNL, *Coyote Series Data Report*, UCID: 19953 Vol. 1, 2. LLNL/NWC, Lawrence Livermore National Laboratory, Berkeley, CA, 1983.
35. LLNL, *Vapor Burn Analysis for the COYOTE series LNG spill Experiments*, UCID: 53530, LLNL/NWC, Lawrence Livermore National Laboratory, Berkeley, CA, 1984.
36. Zhai, X., Ding, S., Cheng, Y., Jin, Y., and Cheng, Y., CFD simulation with detailed chemistry of steam reforming of methane for hydrogen production in an integrated micro-reactor, *International Journal of Hydrogen Energy*, 35, 5383, 2010.
37. Dou, B., and Song, Y., A CFD approach on simulation of hydrogen production from steam reforming of glycerol in a fluidized bed reactor, *International Journal of Hydrogen Energy*, 35, 10271, 2010.
38. Baraldi, D., Kotchourko, A., Lelyakin, A., Yanez, J., Gavrikov, A., Efimenko, A., Verbecke, F., Makarov, D., Molkov., V., and Teodorczyk, A., An inter-comparison exercise on CFD model capabilities to simulate hydrogen deflagrations with pressure relief vents, *International Journal of Hydrogen Energy*, 35, 12381, 2010.
39. Pasman, H.J., and Groothuisen, Th.M., Design of pressure relief vents. In *Loss Prevention in the Process Industries*, edited by C.H. Bushman, Elsevier, New York, 1974, pp. 185–189.
40. Baraldi, D., Heitsch, M., and Wilkening, H., CFD simulations of hydrogen combustion in a simplified EPR containment with CFX and REACFLOW, *Nuclear Engineering and Design*, 237, 1668, 2007.
41. Redlinger, R., DET3D-ACFD tool for simulating hydrogen combustion in nuclear reactor safety, *Nuclear Engineering and Design*, 238, 610, 2008.
42. Dorofeev., S.B., Kuznetsov, M.S., Alekseev, V.I., Efimenko, A.A., and Breitung, W., Evaluation of limits for effective flame acceleration in hydrogen mixtures, *Journal of Loss Prevention in the Process Industries*, 14, 583, 2001.

43. Xiong, J., Yang, Y., and Cheng, X., CFD application to hydrogen risk analysis, *Science and Technology of Nuclear Installations*, 2009, doi: 1155/2009/213981.
44. Heitsch, M., Huhtanen, R., Techy, Z., Fry, C., and Kostka, P., CFD evaluation of hydrogen risk mitigation measures in a VVER-440/213 containment, *Nuclear Engineering and Design*, 240, 385, 2010.
45. Prabhudharwadkar, D.M., and Iyer, K.N., Simulation of hydrogen mitigation in catalytic recombiner, Part II: Formulation of a CFD model, *Nuclear Engineering and Design*, 241, 1758, 2011.
46. Ferng, Y.M., and Chen, C.T., CFD investigating thermal-hydraulic characteristics and hydrogen generation from graphite-water reaction after SG tube rupture in HTR-10 reactor, *Applied Thermal Engineering*, 31, 2430, 2011.

6

Hazards of Hydrogen Use in Vehicles

6.1 Hydrogen Systems in Vehicles

As early as 1978, Professor Trevor Kletz introduced the principles of inherent safety (see Chapter 7 for a detailed presentation). One of his ideas was that *"What you do not have can't leak"* [1]. Nevertheless, as long as we need a source of energy to move our cars, we have to choose among gasoline, diesel fuel, liquid propane gas, natural gas, hydrogen, or some other source. Once we choose hydrogen as a motor fuel, we need to identify principles of inherent safety to cope with the hazards stemming from this new fuel; for instance, minimization of stored quantities and the length of stay in the hazardous zone, especially in refueling stations.

It should be noted also that, although hydrogen has been proven relatively safe in the chemical and aerospace industries, its anticipated widespread application in the automotive industry and consequently, in our everyday lives will necessitate handling of large quantities of hydrogen in gaseous or liquid states by unskilled personnel. This will considerably increase the frequency of small to medium accidents. Obviously, significant attention should be paid to the design of inherently safer hydrogen-fueled cars, safer storage systems onboard, and refueling stations that take into account human error.

Another safety issue is the transport of huge quantities of hydrogen by tankers that have proven quite unsafe on roads as expressed by their high accident rates. Alternatively, pipeline systems look much safer and these are to be used in the near future for the mass transport of hydrogen, either in gaseous or liquid state. Hydrogen pipelines with operating pressures up to 100 MPa have been in use in Europe for many decades with no accident reports. Of these pipelines, the longest ones are found in Germany (215 km) since 1938 and in France (290 km) since 1966 [2].

6.1.1 Internal Combustion Engines

Hydrogen may be used in one of three ways to power vehicles:

- As a replacement for gasoline or diesel fuel in an internal combustion engine,
- As a supplement to gasoline or diesel fuel used in an internal combustion engine, or
- To produce electricity in a fuel cell.

Internal combustion engines based on the Otto and Diesel cycles can be operated with hydrogen or hydrogen mixtures with other liquid fuels. Ricardo (1924) and Burstall (1927) were the first to carry out investigations on using hydrogen as a fuel in vehicles, but it was Erren (1930) who performed intensive studies on hydrogen-air and hydrogen-oxygen motors. The first internal combustion engine car in the United States fueled by hydrogen was presented by Billings in Utah in 1966. Today, significant developments are underway in Germany and Japan [3].

Hydrogen cannot be used directly as a fuel for replacement of gasoline; therefore, engine modifications are necessary. The problematic hydrogen properties to be taken into account are its low ignition energy and high flame propagation speeds that can cause self-ignition during the mixture preparation or flame flashback. The octane number of hydrogen is much lower than that of gasoline and this can cause low performance and fast wearing out of the engine. Nevertheless, the wide ignition range of hydrogen allows burning of lean fuel-air mixtures and this gives a large control range. Uncontrolled premature ignition or even flashback into the intake manifold is usually prevented by adding water as a ballast and also by timed individual port injection of hydrogen close to the cylinder intake. The only noxious gases found in hydrogen-fueled engine emissions are nitric oxides, and many papers have been published on this issue.

6.1.2 Hydrogen Storage in Vehicles

The current possibilities for storage of hydrogen in vehicles are in the gaseous state inside pressure vessels, in the liquid state in vacuum-insulated tanks, or absorbed in metal hydride storage tanks. Unfortunately, none of these hydrogen storage systems offer the irrefutable advantages of gasoline as a fossil energy carrier. Pressurized gas storage vessels are heavy and occupy a large volume on a vehicle. In addition, the high storage pressure of >20.0 MPa raises serious safety concerns.

Liquid hydrogen tanks are larger than gasoline tanks and require complicated construction in order to maintain very low storage temperatures. Safety problems arise in this case too, stemming from the cryogenic conditions and evaporation loss.

Hydride forming metals may be either of the low-temperature type, such as FeTi, with a reversible hydrogen absorption temperature range of 25°C to 100°C, or of the high-temperature hydride material type, such as Mg_2Ni, desorbing hydrogen at much higher temperatures (about 350°C). The latter materials have a higher hydrogen capacity but this comes at the expense of more energy to desorb the stored hydrogen and with increased safety considerations due to a much higher operating temperature.

6.1.3 Safety Comparisons of Hydrogen, Methane, and Gasoline

Substitution of conventional fuels by alternative energy carriers has been implemented to some extent by the introduction of natural gas as a generalized fuel in the world market. Its use is not limited to industry and the home, but extends to public means of transportation, especially in Europe. The prospects for hydrogen use are similar to those of natural gas and the proposal for their combined use has been made. Besides the techno-economical and environmental advantages discussed for hydrogen application, another significant issue is that of the comparative safety between natural gas and hydrogen concerning the application, transport, and storage procedures.

Thermophysical, chemical, and combustion properties of hydrogen, methane, and gasoline are given for comparison in Table 3.1. Of these fuels, gasoline is certainly the easiest and perhaps the safest fuel to store because of its higher boiling point, lower volatility, and narrower flammability and detonability limits. This generalized consideration is based on our previous discussion concerning fire and explosion hazards. Nevertheless, hydrogen and methane (the principal ingredient of natural gas) can also be safely stored using current technology.

Despite its volumetric energy density, hydrogen has the highest energy-to-weight ratio of any fuel. Unfortunately, this weight advantage is usually overshadowed by the high weight of the hydrogen storage tanks and associated equipment. Thus, most hydrogen storage systems designed for transport applications are considerably bulkier and heavier than those used for liquid fuels such as gasoline or diesel. A comparison of the hazards posed by hydrogen, methane, and gasoline follows [4, 5]:

- *Size of molecules.* Since the hydrogen molecule is the smallest of all it will leak through permeable materials, whereas methane and gasoline will not, but the difference in leakage rates is very low. Hydrogen has approximately three times the energy of methane by mass, yet one-third the energy of methane by volume. Therefore, at the same pressure, three times the volume of hydrogen will have the same total energy content as methane. For pinhole-size leaks from high-pressure systems, about three times the volume of hydrogen will leak over methane.
- *Fuel spills.* In the event of a fuel spill, it is expected that a fire hazard will develop most rapidly in the descending order: hydrogen, methane, and gasoline. With regard to fire duration, gasoline fires last

the longest and hydrogen fires are the shortest ones, while all three fuels burn at nearly the same flame temperature. In fact, for spillage of identical liquid fuel volumes, hydrocarbon fires will last 5 to 10 times longer than hydrogen fires.

- *Odorization.* Natural gas is odorized so that leaks can be detected, whereas gasoline is normally odorous. Since natural gas distribution piping exists in so many places and natural gas is piped into buildings and homes, odorization is a prudent although not entirely effective safety measure. Leaks will only be detected if someone is present to smell them and respond. Hydrogen as an industrial gas or fuel cell vehicle fuel is not odorized because sulfur-containing substances (mercaptans) contaminate the catalysts of a fuel cell.
- *Buoyancy.* Hydrogen is 14.5 times lighter than air at normal temperature and pressure, whereas methane is 1.8 times lighter and gasoline vapor is heavier than air. Thus, hydrogen will rise much more quickly causing greater turbulent diffusion, which reduces its concentration below the lower flammability limit (LFL) more rapidly. Moreover, hydrogen diffuses into air approximately 4 times faster than methane and 12 times faster than gasoline, thus causing rapid concentration decrease to safe levels.
- *Energy of explosion.* The energy of explosion values given in Table 3.1 should be considered as the theoretical maxima, and yield factors of 10 percent are considered reasonable for fuel-air explosions. For equivalent volume storage, hydrogen has the least theoretical explosive potential of the three fuels considered, although it has the highest heat of combustion and explosive potential on a mass basis.
- *Flammability and detonability limits.* The broader flammability and detonability limits of hydrogen coupled with its rapid burning velocity render hydrogen a greater explosive threat than methane or gasoline. The LFL for hydrogen and methane is similar (4.0 percent for hydrogen and 5 percent for methane). However, hydrogen has a much wider range between the LFL and the lower detonability limit (LDL) than methane (4.0 to 18 percent for hydrogen vs. only 5 to 5.7 percent for methane). This means that a hydrogen gas concentration of more than three times that of methane is required to produce a detonable mixture. For safety reasons, LFL is used normally in place of LDL for hydrogen, which provides an additional safety factor. A gas concentration in air equal to 25 percent of the LFL for hydrogen corresponds to 1 percent H_2 in air, but 25 percent of the LDL is 4.5 percent hydrogen in air. Therefore, gas detection using LFL gives an earlier warning for a hydrogen detonable mixture than a methane detonable mixture.

- *Ignition energy.* At concentrations up to approximately 10 percent of hydrogen and methane in air, hydrogen has the same ignition energy as methane. As the hydrogen concentration increases toward a stoichiometric mixture of 29 percent hydrogen in air, the ignition energy drops to about one-fourteenth of that for methane and one-twelfth of gasoline. Since we are generally concerned with the prevention of ignitable mixtures, the LFL is the important property. However, the energy levels required for the ignition of these fuels are so low that common ignition sources, such as static electricity discharges from a human body, will ignite any of these fuels in air.
- *Autoignition temperature.* Hydrogen and methane possess unusually high autoignition temperatures (585°C and 540°C, respectively), whereas gasoline with autoignition temperature ranging from 227°C to 477°C appears more hazardous.
- *Deflagrations.* A confined deflagration of hydrogen-air or methane-air will produce a static pressure rise ratio of less than 8:1. Explosion pressures for confined deflagrations of gasoline-air are about 70 to 80 percent of those for hydrogen-air. Unconfined deflagration overpressures are usually <7 kPa. However, a pressure of 3 to 4 kPa is sufficient to cause structural damage to buildings; therefore, unconfined large volume gas-phase explosions can be destructive. Thus, it is apparent that confined deflagrations with up to 8 atm (811 kPa) of explosion pressure can be devastating and even unconfined deflagrations can cause slight to moderate structural damage, and injure people via fire and window-glass shrapnel.
- *Detonations.* Pressure rise ratios of ~15:1 for hydrogen-air or methane-air detonations and a ratio of ~12:1 for a gasoline-air detonation are normally expected. The impulse created by the explosion pressure profile should be taken into account in evaluating explosion damage and in the design of barricades or other structures constructed to mitigate explosion consequences. Hydrogen burning speed (time to peak pressure) is ten times greater than methane. This indicates that a hydrogen detonation will be of much greater severity, yet with shorter positive phase duration and peak explosion overpressure close to that of methane. As a result, material constructions should respond to the same overpressure in a shorter time period.
- *Shrapnel hazard.* This depends on explosion overpressure and with ordinary enclosures (L/D <30) is about the same for hydrogen-air and methane-air and somewhat less severe for gasoline-air mixtures. Nevertheless, in long structures such as tunnels or pipes, hydrogen poses a greater explosion risk than the other two fuels owing to its greater tendency for a deflagration-to-detonation transition (DDT). Thus, hydrogen presents the greatest hazard of shrapnel damage.

- *Radiant heat.* Owing to the heat-absorbing water vapor created during hydrogen combustion and the absence of a carbon combustion reaction, the radiant heat from a hydrogen fire is significantly less than that of a hydrocarbon fire, and this reduces the risk of secondary fires. Combustible materials may actually be placed closer to a hydrogen flame than a methane flame. The reduced radiant heat means less heating of adjacent equipment in case of a major fire and hence a lower probability for a domino effect that leads to escalating damage and losses.
- *Hazardous smoke.* The potential for smoke inhalation damage is judged to be most severe in the descending order: gasoline, methane, and hydrogen fires.
- *Flame visibility.* Unlike visible methane and gasoline flames, hydrogen burns with a near-invisible flame in daylight, although contaminants in air generally add some visibility. Nevertheless, hydrogen flames are visible at night and modern detection equipment can detect them even in daylight.
- *Fire fighting.* Normally, hydrogen and methane fires should be allowed to burn until gas flow is stopped or until liquid spills are consumed because of the potential explosive hazard resulting from extinguishing these fires. However, the fire should be controlled by cooling the storage tanks with water in all situations. Dry chemicals and high-expansion foams can be used to extinguish methane and gasoline fires.

Conclusively, hydrogen has been used and stored safely in industry for quite a long time as compressed gas or liquefied hydrogen, and it seems that metal hydride storage will be equally safe or even safer. Consideration of future hydrogen applications reveals apparently manageable safety problems in the industrial and commercial markets. Although hydrogen safety problems have been efficiently controlled in the industry, additional safety analyses are needed in the transportation and residential fuel markets.

6.2 Accidents Caused by Hydrogen Use in Vehicles

6.2.1 Hydrogen Vehicle Hazards

Hydrogen is a promising fuel gas for transport; therefore, hazards associated with this application should be thoroughly investigated. These hazards should be considered in situations when a vehicle is inoperable, in normal operation, and in collisions.

Potential hazards are commonly related to fire, explosion, or toxicity as in any other case. *Toxicity* hazards can be ignored because neither hydrogen nor its combustion products are toxic, as mentioned earlier. *Fire and explosion* of hydrogen used for transport may come from the fuel storage in the vehicle, the fuel supply lines, or from the fuel cell, if such a system is used. Among them, the fuel cell is the least hazardous, although hydrogen and oxygen in the existing technology are separated by a thin polymer membrane (20 to 30 μm). The initiating event in this case would be the membrane rupture resulting in the combination of hydrogen and oxygen, yet in such an event the fuel cell would lose its potential, which can be easily detected by a control system. The supply lines can then be immediately disconnected. The operating temperature of a fuel cell (60°C to 90°C) is too low to act as a thermal ignition source, but the two reactants may combine on the catalyst surface to create ignition conditions. Nevertheless, the damage potential is limited since the amount of hydrogen in the fuel cell and the fuel supply lines is small.

The greatest damage potential is located in the fuel gas tank because the largest amount of hydrogen is found there. The *failure modes* that should be considered in normal operation, and in collision as well, are the following:

- *Catastrophic rupture of a tank.* This may be the result of a manufacturing defect, abusive handling or stress fracture, puncture by a sharp object, or external fire combined with failure of a pressure relief device to open.
- *Massive leak.* This may be due to faulty operation of a pressure relief valve, chemically induced failure of the tank wall, puncture by a sharp object, or normal operation of a pressure relief valve in case of fire.
- *Slow leak.* Probable reasons may be stress cracking of the tank liner, faulty operation of a pressure relief valve, faulty coupling from the tank to the feed line, or failure of the fuel line connection.

Because a catastrophic rupture is rather unlikely, failure modes resulting in large or slow hydrogen release have been identified experimentally. Countermeasures to avoid the above failure modes could be, for instance:

- *Leak prevention* by a suitable safety design, allowing for shock and vibration tolerance of high-pressure lines.
- *Leak detection* by a proper detector or by adding an odorant in the fuel, which with the existing technology is a problem for fuel cells.
- *Ignition prevention* by automatic disconnection of the battery in an accident, separation of fuel supply lines and electrical systems, and design of proper systems for both active and passive ventilation, such as openings to allow hydrogen to escape upwards.

Based on these failure modes, a detailed risk assessment of the most probable or most severe *hydrogen accident scenarios* was conducted by Swain, Shriber, and Swain [6], including fuel tank fire or explosion in unconfined spaces and in tunnels, fuel line leaks in unconfined spaces, fuel leak in a garage, and refueling station accidents.

These studies have determined that:

- A well-designed hydrogen fuel cell car should be safer than a natural gas or a gasoline car in collisions in open spaces.
- A hydrogen fuel cell car should be nearly as safe as a natural gas car, and both should be safer than a gasoline or propane car in a tunnel collision.
- The greatest risk appears in the scenario of a hydrogen leak in a garage, which in the absence of passive or active ventilation may result in fire or explosion.

Conclusively, hydrogen fuel presents similarities and differences from other transportation fuels, with some parameters tending to make accidents more severe and others making accidents less severe. Thus, it is not clear whether hydrogen would entail more or less risk in transportation. The good safety records of trucks carrying compressed and liquefied hydrogen over the road, for instance in the United States and Germany, add confidence that there are not large, unknown risks associated with hydrogen fuel use. Despite public perception, the use of hydrogen as a fuel for transport may be safer than gasoline or natural gas in some aspects [7].

6.2.2 Hydrogen Tanks Onboard

In the United States, the Motor Vehicle Fire Research Institute (MVFRI) has contracted with Southwest Research Institute (SwRI) to perform testing of fuel systems. Zalosh, of the MVFRI, conducted a study on CNG (compressed natural gas) and hydrogen vehicle fuel tank failure incidents, in which he gives useful recommendations for both fuels when stored onboard in cylindrical pressure vessels. [8]

The following types of cylinders have been used to store these gases under pressure:

- A type 1 container is a metallic noncomposite container.
- A type 2 container is a metallic liner over which an overwrap, such as carbon fiber or fiberglass, is applied in a hoop-wrapped pattern over the liner's cylinder sidewall.
- A type 3 container is a metallic (usually aluminum) liner over which an overwrap, such as carbon fiber or fiberglass, is applied in a full-wrapped pattern over the entire liner, including the domes.

- A type 4 container is a nonmetallic liner over which an overwrap, such as carbon fiber or fiberglass, is applied in a full-wrapped pattern over the entire liner, including the domes.

Because no relevant hydrogen incidents had been reported at the time of his study, Zalosh suggested that hydrogen tanks behave with the same failure modes as CNG tanks. The MVFRI funded two tests to determine the consequences of fire on these tanks. Descriptions of those tests have been reported by Zalosh and Weyandt [9, 10, 11].

In these experiments propane burner fires were set under cylinders that were filled with hydrogen at a pressure of 32 to 34 MPa without pressure relief devices. The first test was done with a 72 liter type 4 cylinder (Figure 6.1), while the second one used an 88 liter type 3 cylinder installed under a sports utility vehicle (SUV) (Figure 6.2). The type 4 cylinder ruptured after 6 minutes and 27 seconds of fire exposure. The burner flame ignited the SUV before the type 3 cylinder ruptured after 12 minutes and 18 seconds of fire exposure. In both cases, the cylinder wraps prevented the hydrogen temperature and pressure from increasing significantly above their pretest values.

The type 4 hydrogen cylinder primary remains were found about 82 m away from the burner (Figure 6.3), whereas the largest fragment of the type 3 hydrogen cylinder was found about 41 m away from the burner (Figure 6.4).

In these tests, some blast overpressure measurements were performed as shown in Figure 6.5. The ideal blast wave pressures calculated for blast

FIGURE 6.1
Type-4 hydrogen fuel tank test setup on propane burner. (From Zalosh, R., Proceedings of the 5th International Seminar on Fire and Explosion Hazards, Edinburgh, UK, April 23–27, 2007. With permission from the School of Engineering, University of Edinburgh, http://www.see.ed.ac.uk/feh5/pdfs/FEH_pdf_pp149.pdf.)

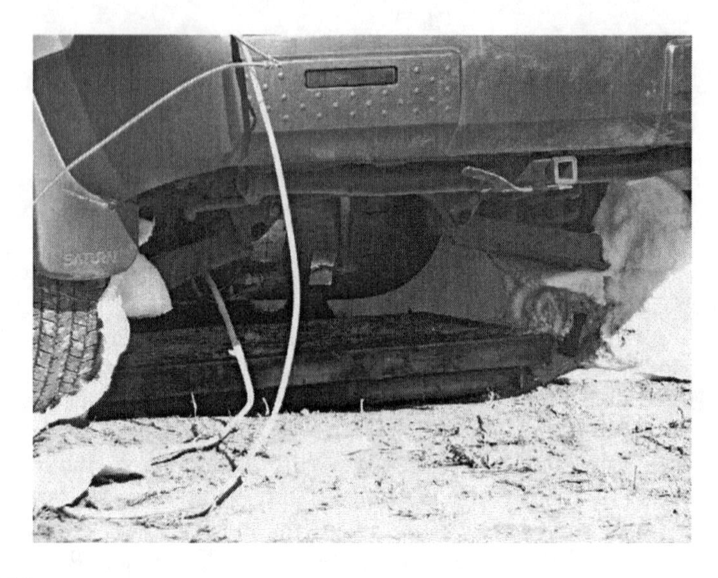

FIGURE 6.2
Burner below type-4 cylinder under an SUV. (From Zalosh, R., Proceedings of the 5th International Seminar on Fire and Explosion Hazards, Edinburgh, UK, April 23–27, 2007. With permission from the School of Engineering, University of Edinburgh, http://www.see.ed.ac.uk/feh5/pdfs/FEH_pdf_pp149.pdf.)

FIGURE 6.3
Type-4 hydrogen cylinder fragment. (From Zalosh, R., Proceedings of the 5th International Seminar on Fire and Explosion Hazards, Edinburgh, UK, April 23–27, 2007. With permission from the School of Engineering, University of Edinburgh, http://www.see.ed.ac.uk/feh5/pdfs/FEH_pdf_pp149.pdf.)

FIGURE 6.4
Type-3 hydrogen cylinder fragment. (From Zalosh, R., Proceedings of the 5th International Seminar on Fire and Explosion Hazards, Edinburgh, UK, April 23–27, 2007. With permission from the School of Engineering, University of Edinburgh, http://www.see.ed.ac.uk/feh5/pdfs/FEH_pdf_pp149.pdf.)

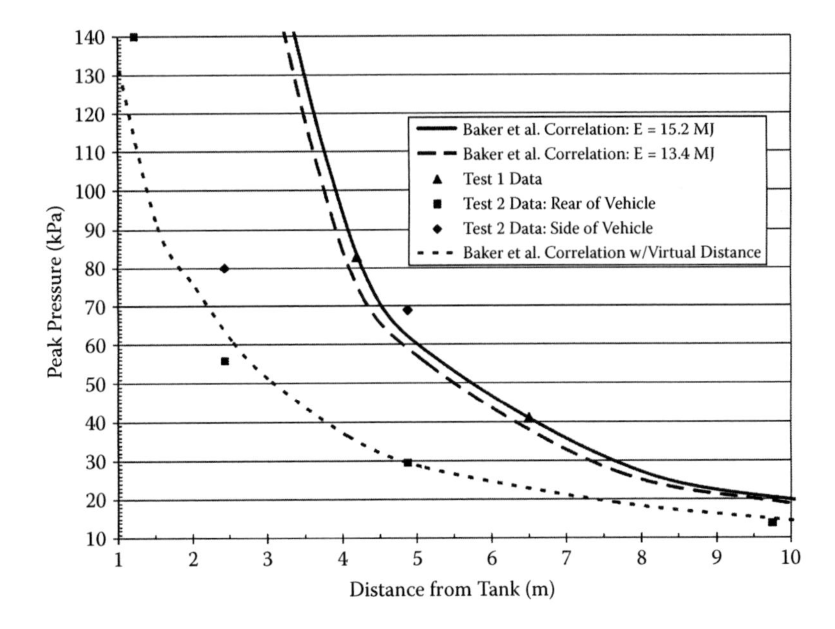

FIGURE 6.5
Pressures of fuel tank blast waves versus distance from tank. (From Zalosh, R., Proceedings of the 5th International Seminar on Fire and Explosion Hazards, Edinburgh, UK, April 23–27, 2007. With permission from the School of Engineering, University of Edinburgh, http://www.see.ed.ac.uk/feh5/pdfs/FEH_pdf_pp149.pdf.)

FIGURE 6.6 (See color insert.)
Hydrogen fireball about 170 m/sec after tank rupture in Test 2 (cylinder installed under an SUV). (From Zalosh, R., Proceedings of the 5th International Seminar on Fire and Explosion Hazards, Edinburgh, UK, April 23–27, 2007. With permission from the School of Engineering, University of Edinburgh, http://www.see.ed.ac.uk/feh5/pdfs/FEH_pdf_pp149.pdf.)

energies of 13.4 MJ and 15.2 MJ are also shown for comparison. The blast energies were calculated by Zalosh [11] from the hydrogen isothermal expansion of the gas inside the pressurized tank during rupture, based on the methodology of Baker et al. [12, 13]. The agreement with the data demonstrates the validity of blast wave calculations based on the cylinder pressure and volume.

In addition to blast waves, fireballs were formed in both tests after the cylinders ruptured (Figure 6.6). In the cylinder-only test, the maximum diameter of the fireball was 7.7 m, whereas, in the cylinder-under-SUV test, the maximum diameter of the fireball was about 24 m. The larger fireball diameter in the SUV test indicates that with respect to formation of the fireball, not only the cylinder rupture, but also the SUV fire contributed to the fireball size.

Conclusively, the results of these tests suggest that the stand-off zone in an anticipated hydrogen tank explosion extends to a radius of more than 100 m from the burning vehicle.

Stephenson [14] investigated the following principal crash-induced fire safety issues related to hydrogen-fueled vehicles:

- Crash forces and exposure to fire
- Fuel tanks
- Fuel lines

- Hydride devices
- Reformers
- Fuel cells
- Regulator failures
- Venting from various sources
- Mechanical energy from tank rupture

Aiming at a reasonable mass fraction of a vehicle, the compressed hydrogen tank should preferably be a type 3 (aluminum liner) or type 4 (plastic liner) carbon fiber-wrapped tank. Typical pressures in these tanks are 345 to 690 bar (34.5 to 69 MPa). Although these pressures seem rather hazardous, the tanks are so well constructed as to withstand that much pressure in a way that they are unlikely to burst in a crash. On the contrary, the tanks may be more vulnerable at low pressures when the tank wall is less supported from the inside.

With regard to *regulators*, many suppliers construct regulators that are placed at least partially inside the tank. In this way, the regulators are well protected and it is quite unlikely that they will be torn off in a crash. This event would cause a more severe accident owing to the rapid evacuation of the entire contents of the tank and its ignition either by existing ignition sources or by autoignition due to static electricity discharges. A countermeasure to avoid such high-pressure hydrogen releases is to have an in-tank solenoid shut-off valve that will stop as soon as possible the high-pressure hydrogen release.

A worst-case scenario is the exposure of the hydrogen tank to a liquid hydrocarbon *pool fire* from another vehicle following a crash. In this case a fiber composite tank is a good insulator that may prevent the temperature and pressure from rising considerably. Although a partially filled tank may not exceed its normal operating pressure, the fire will gradually weaken the carbon fiber or fiberglass overwrap and eventually the tank will burst if the hydrogen is not vented. This is usually accomplished by a thermally activated pressure relief device (PRD), provided that the relief valve receives the same heat flux from the fire as the tank. Consequently, the use of both thermally activated and pressure activated PRDs would be preferable.

Another issue is the *mechanical energy* stored in the compressed gas tank which could see an impulse of about 6000 Newton-seconds if a tank ruptures and leaks and produces a horizontal jet in a preferred direction. This impulse would accelerate a 1360 kg car on a friction-free surface to a velocity of about 16 km per hour [14].

Aging of tanks is also a severe issue. Thus, all compressed gas tanks should have an expiration date to protect against fatigue and corrosion failures (for instance, at the most 15 years of tank life). A high-pressure regulator and several meters of tubing are needed to lower the supply pressure to about

10 bar (1 MPa). Since all of these components may be severely damaged in a crash and with ignition sources nearby, it is reasonable to assume that leaking hydrogen will ignite. Consequently, keeping the trapped volumes low so that the amount of energy release from such small fires is less than that required to ignite nearby materials is indispensable. Furthermore, materials in the vicinity of hydrogen tubes should be carefully selected, accompanied by wise spacing as well.

Sometimes, a *hydride storage device* of some kind may be part of the system. If this device is housed in a pressure vessel, this should hold the full pressure in the event the whole quantity of hydrogen is released from the hydride; thus, a pressure relief device is necessary in this case. Furthermore, all of the hydrogen should be driven out, if the hydride system were to be exposed to an electrical fire or liquid hydrocarbon pool fire.

Fuel cells also contain small quantities of hydrogen that could easily cross the thin membrane and escape in the event of a crash; hence, it will be necessary to safely vent the releasing hydrogen. *Reformer systems* also have some inventory of flammable and hot gases, careful management of which is needed.

In summary, the most important countermeasures according to Stephenson [14] include:

1. The location and protection of major components and routing of electrical wires and fuel lines.
2. The selection of low flammability materials for parts that might be exposed to minor electrical or hydrogen fires.
3. The rapid disconnection of electrical and hydrogen sources after detection by the vehicle crash sensors of crashes over a specified severity. The hydrogen can also be shut off via a variety of system sensors, such as low or high pressures or temperatures, in various lines or components.

6.2.3 Accidental Release of Hydrogen from Vehicles

With regard to the safety concerns raised on the use of hydrogen as a motor fuel, Swain [15] produced a video demonstrating the characteristic features of a fuel leak and ignition for a hydrogen-fueled car and a gasoline-fueled car in open air. Figure 6.7 shows some of the well-known footage of a hydrogen car fire and a gasoline car fire in open air one minute after ignition. It is evident from this photo that the hydrogen can generate very long vertical jet flames and high flame temperature. Nevertheless, the body of the hydrogen-fueled car was not ignited, and the flame lasted only 100 seconds. It is noteworthy that the maximum temperature inside the back window was only 19.5°C. In contrast, the leak from the gasoline-fueled car formed a pool fire, which engulfed the whole car. The car continued to burn for several minutes and was completely destroyed. This demonstration showed that a

FIGURE 6.7
Photo from a video that compares fires from an intentionally ignited hydrogen tank release to a small gasoline fuel line leak. Hydrogen-fueled car (left), gasoline-fueled car (right). Time: 1 min after ignition when hydrogen flow starts to subside while gasoline vehicle fire begins to enlarge. After 100 seconds, all of the hydrogen was gone and the car's interior was undamaged. (From Swain, M.R., *Proceedings of the 2001 DOE Hydrogen Program Review*, NREL/CP-570-30535, U.S. Department of Energy, Washington, D.C., http://www1.eere.energy.gov/hydrogenandfuelcells/pdfs/30535be.pdf.)

hydrogen-fueled car is less hazardous than a gasoline-fueled car in open air, at least under the specific test conditions. Yet, the consequences of a hydrogen-fueled vehicle fire would be much more severe inside a semi-confined space such as a road tunnel, or even worse in an enclosed space such as a garage. This is why a great many studies have been conducted recently dealing with this issue, as subsequently discussed here and also in Chapter 9.

Underground tunnels play an important role in modern road transportation systems. Thus, in a hydrogen economy, hydrogen cars will regularly circulate in ordinary underground tunnels. Wu [16] investigated potential fire hazards and fire scenarios of hydrogen cars in road tunnels, as well as the implications for fire safety measures and ventilation systems in existing tunnels. The CFD simulations carried out by this author concerned two fire scenarios showing that the jet flame hazard can be unique for hydrogen cars. For high release rate, the flame might cause an oxygen deficit inside the tunnel. Impingement of the hydrogen jet flame on the tunnel ceiling would produce high ceiling temperatures causing damage to tunnel infrastructures. In addition, the oxygen deficit caused by the hydrogen fire may also pose a flash fire hazard inside the tunnel and ventilation ducts. In Figure 6.8, the smoke flow in the tunnel controlled by the longitudinal ventilation is shown. The upstream smoke layer is called "backlayering" and is sensitive to ventilation velocity. The "critical velocity" is the velocity that just about

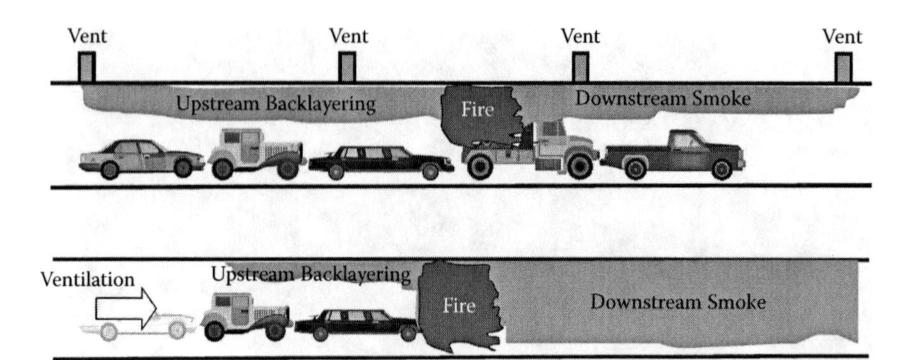

FIGURE 6.8
A tunnel fire and smoke flow under the influence of longitudinal ventilation. (From Wu, Y., *Transportation Research Part* C, 16, 246, 2008. With permission from Elsevier.)

completely eliminates the upstream backlayering and prevents smoke flow from traveling against the ventilation, thus forcing the smoke to move in only one direction.

CFD modeling of hydrogen release, dispersion, and combustion for various automotive scenarios was performed by Venetsanos et al. [17], aiming at investigating the potential effects of releases from compressed gaseous hydrogen systems on commercial vehicles in urban and tunnel environments. The authors also performed comparative releases from compressed natural gas systems. Typical nonarticulated single-deck city buses were analyzed in this study. The analysis showed that in worst-case scenarios in tunnels a deflagration may occur, which, in very unfavorable conditions causing intense turbulence (e.g., multiple obstacles), may evolve to a deflagration-to-detonation event.

A CFD simulation study to investigate the risk from hydrogen-fueled vehicles inside tunnels was performed by Middha and Hansen [18]. The study concerned cars containing 700-bar gas tanks releasing either upward or downward, or liquid hydrogen tanks releasing only upward. Buses were also studied containing 350-bar gas tanks and releasing upward in two different tunnel layouts and a range of longitudinal ventilation conditions. A probabilistic risk assessment study was also performed resulting in a maximum pressure level of 0.1 to 0.3 bar(g), which represents a limited human fatality risk, yet with somewhat higher pressures in specific places (e.g., under the vehicles) due to wave enhancement from reflections.

Baraldi et al. [19] investigated the scenario of a large-scale accidental release and subsequent ignition and deflagration of hydrogen in a 78.5 m long tunnel using numerical analysis in their simulation. Experiments have shown that explosions in such semiconfined spaces can result in severe consequences due to the entrapment of hydrogen for some time inside the structure; this increases the probability of ignition and also leads to the enhancement of pressure waves caused by the tunneling effect. Thus, 1 kg of hydrogen at stoichiometric ratio generated an overpressure peak of about 150 kPa in a tunnel environment, whereas the same quantity and composition resulted in only 10 kPa when ignited in the open. It is noteworthy that the presence of vehicles inside the tunnel did not increase considerably the peak pressure. Numerical simulations using five different CFD codes were partly successful with regard to pressure peak prediction, whereas the agreement between experimental data and simulation results was not satisfactory for the pressure rise rate. This discrepancy was attributed to the mesh resolution, the numerical scheme, and some inaccuracy in the physical description of the flame acceleration.

A potential risk associated with the use of hydrogen in automotive applications is a slow, long-lasting hydrogen release from a vehicle's components into an inadequately ventilated enclosed structure such as a garage. This leak may stem from permeation through the containment materials found in compressed gaseous hydrogen storage systems and facilitated by the extremely small size of hydrogen molecules. Consequently, this is an issue that requires special consideration for containers, especially when constructed of nonmetallic liners (mainly polymers). The molecular diffusion through the walls or interstices of a container vessel, piping, or interface material has been established by SAE International [20] and cited by Makarov and Molkov [21] as shown in Table 6.1.

Adams et al. [22] investigated the same problem and developed a methodology to estimate an allowable upper limit for the hydrogen permeation rate from road vehicles into enclosed spaces. After this study, the worst garage ventilation rate was determined with SAE to be 0.03 ac/h (air changes per hour) and the maximum prolonged material temperature as 55°C. In addition, the authors concluded that stratification inside these spaces is insignificant and, therefore, it would be valid to assume homogeneous distribution of hydrogen at the flow and ventilation rates considered. They have also concluded that since the effects of aging on the permeation behavior of complete containers is uncertain, a factor of two should be used for calculating for life time and replacing containers.

TABLE 6.1

Comparative Permeability of Materials to Hydrogen $[(\text{mol } H_2) \cdot s^{-1} \cdot m^{-1} \cdot MPa^{-1/2}]$

Material	T = 20°C	T = 55°C
Iron	5.47×10^{-11}	2.38×10^{-10}
High density polyethylene (HDPE)	9.30×10^{-13}	3.17×10^{-12}
Carbon fiber/epoxy composite	1.85×10^{-13}	$5.79\ 10^{-13}$
Stainless steel 303	4.09×10^{-16}	7.69×10^{-15}
Aluminum	$2.47\ 10^{-24}$	7.03×10^{-22}

Source: Data from References 20 and 21.

6.2.4 Fueling Stations

Although the first commercial hydrogen-fueled cars are now available in the marketplace (although at very high prices), the principal problem for using such a car today is the absence of a network of refueling stations. Thus, the owner of a hydrogen-fueled vehicle must limit travel to distances where the vehicle can be refueled securely; clearly, the lack of hydrogen infrastructure significantly hinders the emergence of the so-called hydrogen economy. The declarations for a national policy to be prepared for the hydrogen economy, first by the prime minister of Iceland in 2001 and shortly thereafter by the former European Commission President Romano Prodi in 2002, were very promising and the efforts toward this aim are substantial. Then, in January 2003, President George W. Bush shared the same sentiment in his State of the Union Address. In January 2004, Governor Arnold Schwarzenegger declared in his State of the State Address that California would have a hydrogen highway [23]. Consequently, the safety of hydrogen refueling stations should be investigated in parallel with other hydrogen safety issues.

Markert et al. [2] analyzed a number of safety concerns related to various potential future hydrogen infrastructures to be built for a sufficient supply of refueling stations within an urban region. The authors suggest the need for a rapid change of building codes that concern specifically the use of fuels giving off vapors heavier than air, such as gasoline vapor and LPG, but not for hydrogen, which is much lighter than air. For instance, it is required that ventilation openings in garages should be near the ground with often no openings at higher levels. Evidently, such enclosed spaces are prone to destructive hydrogen explosions.

The safety of hydrogen refueling stations could be increased by reducing the amount stored at the station according to inherently safer design principles (Chapter 7). This could be accomplished by pipeline transmission or even by small on-site hydrogen production facilities. In the case of truck delivery, a measure to increase safety could be just leaving the trailer at the refueling station instead of unloading the tankers into buffer storage. With regard to cars, refueling safety could be greatly improved by exchanging whole car tanks at refueling stations, which could be easily and safely recharged as cartridges at remote sites.

Kikukawa [24] conducted a consequence analysis and safety verification of hydrogen fueling stations using CFD simulation. The study was validated against the results of horizontal squirt tests of hydrogen leaking from a pinhole (0.2 mm) in high-pressure gas facilities, such as the piping in a hydrogen refueling station. Ignition by both static electricity (autoignition) and a flame outside the station were analyzed.

Baraldi et al. [25] performed CFD simulations of hydrogen dispersion and combustion in accident scenarios in a mock-up liquid hydrogen refueling station, investigating accident scenarios caused by a hose break during LH_2 vehicle refueling. Both a nonmitigated and a mitigated accident scenario

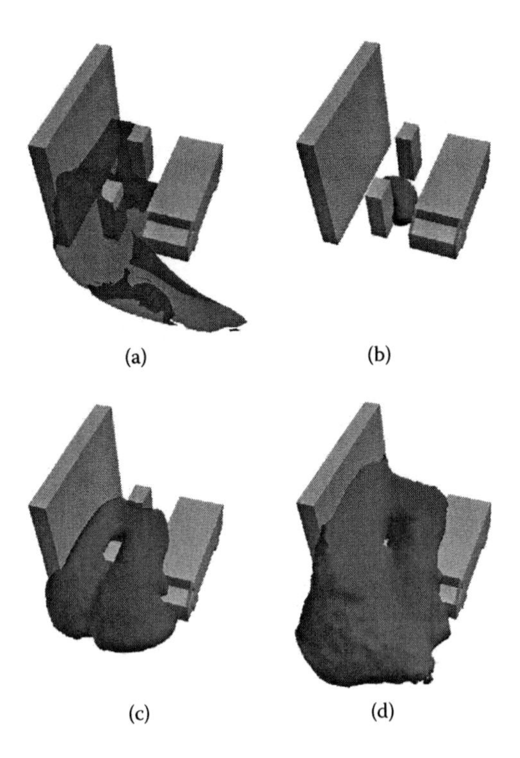

FIGURE 6.9
Non-mitigated west wind scenario. Ignition is located between the dispensers. The molar fraction (0.04) in blue color (a) shows the initial hydrogen flammable cloud. Isosurface of temperature (1000 K) in red color (b,c,d) shows the propagation of the flame [25] (With permission from Elsevier.)

were analyzed assuming five different ambient conditions. It was found that the expectation that the wind plays a positive role in increasing the dispersion of the flammable cloud, thus reducing the overpressures in eventual explosions, is not justified for all wind directions. The propagation of the flame for the nonmitigated scenario with the wind coming from the west is shown in Figure 6.9.

Toward the same goal and utilizing the same mock-up hydrogen refueling station, Makarov et al. [26] simulated a hydrogen explosion involving hydrogen stored in a gaseous state. A total of seven partners obtained nine simulation results using various CFD tools. The paper details the models and numerical codes used and presents the simulated pressure transients obtained in comparison with the experimental pressure records. The comparative model analysis was based on simulation results achieved. The simulated maximum overpressure and the characteristic rate of pressure rise were treated as major output parameters. The simulations had, in general, a reasonable agreement with experimental data, although some codes predicted relatively large overpressures in the recirculation area behind the car.

Quantitative risk assessment (QRA) of a predefined hypothetical gaseous hydrogen refueling station was investigated by Papanikolaou et al. [27]. In total, 15 scenarios were simulated. Five of them concerned hydrogen releases in confined ventilated spaces (inside the compression and the purification/drying buildings) and 10 scenarios concerned releases in open and semiconfined spaces (in the storage cabinet, storage bank, and refueling hose of one dispenser). The simulations of the open and semiconfined scenarios showed that the wind velocity has an insignificant effect on the size of the flammable clouds. The limited effect of the wind velocity is attributed to the presence of the surrounding buildings that cause a wind velocity field within the refueling station. The risk assessment parameters (such as maximum flammable hydrogen mass and mixture volume and maximum horizontal and vertical distance of LFL cloud) are mainly influenced as expected by the release rate (i.e., the leak diameter size).

A benchmarking exercise on QRA methodologies for hydrogen safety in a virtual refueling station was conducted by Ham et al. [28] (see also Chapter 9). The benchmark exercise showed that big differences exist in the approach of QRA and in the nature of the results obtained, probably owing to limitations in the codes used. The scatter is wider in the computations of dispersion parameters (e.g., dimension of flammable cloud and explosive mass). Nevertheless, this scatter does not appear in a clustering of numerical results versus analytical results.

References

1. Kletz, T.A., What you don't have, can't leak. *Chemistry and Industry*, May 1978, 287–292.
2. Markert, F., Nielsen, S.K., Paulsen, J.L., and Andersen, V., Safety aspects of future infrastructured scenarios with hydrogen refuelling stations, *International Journal of Hydrogen Energy*, 32, 2227, 2007.
3. U.S. Department of Transportation, *Guidelines for Use of Hydrogen Fuel in Commercial Vehicles*, Federal Motor Carrier Safety Administration, November 2007.
4. Rigas, F., and Sklavounos, S., Evaluation of hazards associated with hydrogen storage facilities, *International Journal of Hydrogen Energy*, 30, 1501, 2005.
5. Hord, J., Is hydrogen a safe fuel? *International Journal of Hydrogen Energy*, 3, 157, 1978.
6. Swain, M.R., Shriber, J., and Swain, M.N., Comparison of hydrogen, natural gas, liquefied petroleum gas, and gasoline leakage in a residential garage, *Energy and Fuels*, 12, 83, 1998.
7. Farrell, A.E., Keith, D.W., and Corbett, J.J., A strategy for introducing hydrogen into transportation, *Energy Policy*, 31, 1357, 2003.

8. Zalosh, R., *CNG and Hydrogen Vehicle Fuel Tank Failure Incidents. Testing, and Preventive Measures, Technical Support and Evaluation of the Fuel Tank Tests; Recommendations for Research Priorities of Hydrogen Fueled Vehicles*, Motor Vehicle Fire Research Institute, 2008.
9. Zalosh, R., and Weyandt, N., "Hydrogen Fuel Tank Fire Exposure Burst Test," SAE Paper No. 2005-01-1886, 2005.
10. Weyandt, N., "Intentional Failure of a 5000 psig Hydrogen Cylinder Installed in an SUV without Standard Required Safety Devices," SAE Paper No. 2007-01-0431, 2007.
11. Zalosh, R., "Blast Waves and Fireballs Generated by Hydrogen Fuel Tank Rupture During Fire Exposure," Proceedings of the 5th International Seminar on Fire and Explosion Hazards, Edinburgh, UK, April 23–27, 2007.
12. Baker, W., Kulesz, J., Ricker, R., Westine, P., Parr, V., Vargas, L., and Mosely, P., "Workbook for Estimating Effects of Accidental Explosions in Propellant Ground Handling and Transport Systems," NASA CR 3023, August 1978.
13. Center for Chemical Process Safety, *Guidelines for Chemical Process Quantitative Risk Analysis*, 2nd edition, American Institute of Chemical Engineers, New York, 2000.
14. Stephenson, R., "Crash-induced Fire Safety Issues with Hydrogen-Fueled Vehicles," Presented at National Hydrogen Association's 18th Annual U.S. Hydrogen Conference, Washington, D.C., March 2003.
15. Swain, M.R., Fuel leak simulation, in *Proceedings of the 2001 DOE Hydrogen Program Review*, NREL/CP-570-30535, U.S. Department of Energy, Washington, D.C.
16. Wu, Y., Assessment of the impact of jet flame hazard from hydrogen cars in road tunnels, *Transportation Research Part C*, 16, 246, 2008.
17. Venetsanos, A.G., Baraldi, D., Adams, P., Heggem, P.S., and Wilkening, H., CFD modeling of hydrogen release, dispersion and combustion for automotive scenarios, *Journal of Loss Prevention in the Process Industries*, 21, 162, 2008.
18. Middha, P., and Hansen, O.R., CFD simulation study to investigate the risk from hydrogen vehicles in tunnels, *International Journal of Hydrogen Energy*, 34, 5875, 2009.
19. Baraldi, D., Kotchourko, A., Lelyakin, A., Yanez, J., Middha, P., Hansen, O.R., Gavrikov, A., Efimenko, A., Verbecke, F., Makarov, D., and Molkov, V., An intercomparison exercise on CFD model capabilities to simulate hydrogen deflagrations in a tunnel, *International Journal of Hydrogen Energy*, 34, 7862, 2009.
20. SAE International, Technical information report J2579, January 2009.
21. Makarov, D., and Molkov, V., "Modelling of dispersion following hydrogen permeation for safety engineering and risk assessment," Presented at II International Conference: "Hydrogen Storage Technologies," Moscow, Russia, October 28–29, 2009.
22. Adams, P., Bengaouer, A., Cariteau, B., Molkov, V., and Venetsanos, A.G., Allowable hydrogen permeation rate from road vehicles, *International Journal of Hydrogen Energy*, 36, 2742, 2011.
23. Clark, W.W., Rifkin, J., O'Connor, T., Swisher, J., Lipman, T., and Rambach, G., Hydrogen energy stations: Along the roadside to the hydrogen economy, *Utilities Policy*, 13, 41, 2005.
24. Kikukawa, S., Consequence analysis and safety verification of hydrogen fueling stations using CFD simulation, *International Journal of Hydrogen Energy*, 33, 1425, 2008.

25. Baraldi, D., Venetsanos, A.G., Papanikolaou, E., Heitsch, M., and Dallas, V., Numerical analysis of release, dispersion and combustion of liquid hydrogen in a mock-up hydrogen refuelling station, *Journal of Loss Prevention in the Process Industries*, 22, 303, 2009.
26. Makarov, D., Verbecke, F., Molkov, V., Roe, O., Skotenne, M., Kotchourko, A., Lelyakin, A., Yanez, J., Hansen, O., Middha, P., Ledin, S., Baraldi, D., Heitsch, M., Efimenko, A., and Gavrikov, A., An inter-comparison exercise on CFD model capabilities to predict a hydrogen explosion in a simulated vehicle refuelling environment, *International Journal of Hydrogen Energy*, 34, 2800, 2009.
27. Papanikolaou, E., Venetsanos, A.G., Schiavetti, M., Marangon, A., Carcassi, M., and Markatos, N., Consequence assessment of the BBC H_2 refuelling station using the ADREA-HF code, *International Journal of Hydrogen Energy*, 36, 2573, 2011.
28. Ham, K., Marangon, A., Middha, P., Versloot, N., Rosmuller, N., Carcassi., M., Hansen, O.R., Schiavetti., M., Papanikolaou, E., Venetsanos., A., Engebo, A., Saw., J.L., Saffers., J.B., Flores, A., and Serbanescu, D., Benchmark exercise on risk assessment methods applied to a virtual hydrogen refueling station, *International Journal of Hydrogen Energy*, 36, 2666, 2011.

7

Inherently Safer Design

As discussed by Amyotte, MacDonald, and Khan [1], inherent safety is a proactive approach in which hazards are eliminated or lessened so as to reduce risk without over-reliance on engineered (add-on) devices and procedural measures. The concepts of inherent safety (or *inherently safer design*, ISD) have been formalized in the process industries over the past 35 years, beginning with the pioneering work of Professor Trevor Kletz (largely in response to the cyclohexane explosion at Flixborough, in the United Kingdom, in 1974). Many publications on ISD are now available: recent texts [2, 3], review articles including Khan and Amyotte [4], and resource articles such as Hendershot [5].

Professor Kletz and others worldwide have formulated a number of principles or guidelines to facilitate inherent safety implementation in industry. Four basic principles have gained widespread acceptance: (1) minimization (or intensification), (2) substitution, (3) moderation (or attenuation), and (4) simplification.

Minimization calls for the use of smaller quantities of hazardous materials when the use of such materials cannot be avoided. It may also involve performing a hazardous procedure as few times as possible when the procedure is unavoidable. *Substitution* calls for the replacement of a substance with a less hazardous material, or a processing route with one that does not involve hazardous material. It could also involve replacing a hazardous procedure with one that is less hazardous. *Moderation* implies the use of hazardous materials in their least hazardous forms, or the identification of processing options that involve less severe conditions, for example, a lower temperature, pressure, or speed of rotation. *Simplification* requires the design of processes, processing equipment, and procedures in a manner so as to eliminate opportunities for errors by eliminating excessive use of add-on safety features and protective devices.

These descriptions indicate that the principles of ISD focus on the underlying chemistry and physics of materials and processing methodologies. In doing so, they directly address the issue of material hazards and the ensuing risk. This approach is preferable to one in which an attempt is made to assess whether a given energy alternative such as hydrogen is *safe* [6]; hence the use of the term inherently *safer* design. ISD concepts, in conjunction with appropriate safety systems including detection, have been recommended as a good starting point for dealing with aspects of hydrogen safety [7].

The chemical and physical properties of hydrogen present both advantages and disadvantages with respect to general safety considerations and those

TABLE 7.1

Safety Advantages and Disadvantages of Hydrogen

Advantages	Disadvantages
Leaks typically disperse at a high rate because of buoyancy effects.	Leaks cannot be identified by senses of sight or smell.
Health effects due to toxicity are not evident (although asphyxiation concern remains).	Ignition energy is low and readily attainable in industry.
Pooling does not occur.	Range between lower and upper flammability limits is wide (and lower flammability limit is relatively low).
	Flames are not visible.
	Significant overpressures can result from explosion.

Source: Adapted from Miller, M., *Hydrogen Fueling Stations*, Institute of Transportation Studies, November 15, 2004.

related specifically to inherent safety. Safety bulletins on hydrogen (e.g., [8] and [9]) typically comment on the fact that it is colorless, odorless, and tasteless, meaning that most human senses will not help to detect a leak [9]. On the other hand, and with possible safety benefits, hydrogen is lighter than air, diffuses rapidly, and is nontoxic and nonpoisonous [9].

Molkov [10] has demonstrated that the ubiquitous safety concept of *trade-offs* applies to hydrogen as well as hydrocarbons. On the safety asset side, he states that buoyancy—such that the formation of large combustible clouds may be less likely with hydrogen than with heavier hydrocarbons—is the main advantage. Negative aspects include the ability of a hydrogen-air mixture to detonate (which is greater than that of hydrocarbons) and the higher molecular diffusion of hydrogen compared to other fuels [10]. Miller [11] has summarized the safety advantages and disadvantages of hydrogen based on its inherent properties, as shown in Table 7.1.

Examples of application of the various principles of inherent safety to the hydrogen industry are given in this chapter; general material on ISD is drawn from Kletz and Amyotte [3], with relevant excerpts. Attempts to measure the degree or level of inherent safety in a given process are also described, both generally and with specific emphasis on those measurement techniques designed for use with hydrogen. First, the concept of an ordered approach to risk reduction is presented in the following section.

7.1 Hierarchy of Risk Controls

The principles of inherent safety work in conjunction with other means of reducing risk, namely, *passive* and *active engineered safety*, and *procedural safety*, within a framework commonly known as the *hierarchy of controls* (or as some call it, the *priority of controls* or the *safety decision hierarchy*) [1, 3]. Inherent safety, being the most effective and robust approach to risk reduction, sits at the top of the hierarchy; it is followed in order of decreasing effectiveness by passive engineered safety devices (such as explosion relief vents), then active engineered safety devices (such as automatic fire suppression systems), and finally procedural safety measures (such as ignition source control by hot-work permitting). Figure 7.1 gives a schematic representation of this way of thinking about loss prevention.

Inherent safety is not a stand-alone concept [1]; as just described and as shown in Figure 7.1, ISD works through a hierarchical arrangement in concert with engineered and procedural safety to reduce risk. However, inherent safety is not necessarily a solution for all hazards and all risks. Nor does the hierarchy of controls invalidate the usefulness of engineered and procedural safety measures. Quite the opposite; the hierarchy of controls recognizes the importance of engineered and procedural safety by highlighting the need for careful examination of the reliability of both mechanical devices and human actions. For example, Molkov [10] comments that while venting is the most widespread mitigation technology, its application to confined hydrogen-air deflagrations can possibly enable a transition to detonation to occur, resulting in greatly increased overpressures. This, of course, is the opposite of the intended effect of lessening the severity of such an incident [10].

As illustrated in Figure 7.2, Hendershot [5] described the hierarchy of chemical process safety strategies as a *spectrum of options*. These considerations are readily apparent in the article by Pasman and Rogers [12], which gives several examples of inherent safety features (as discussed in subsequent sections of this chapter), along with passive devices (e.g., fire walls and explosion barriers), active devices (e.g., sensors, blocking valves, ventilation, and recombiners that catalytically oxidize hydrogen), and procedural measures (e.g., emergency response).

Not all hydrogen safety publications recognize the importance of ISD and some seemingly attempt to invert the risk control hierarchy by promoting safety procedures and training as "probably the most important preventive measure" [13]. Others clearly recognize the importance of inherent safety by emphasizing that "the main focus here should be to avoid significant flammable gas clouds" [14]; this publication also implicitly adopts the hierarchy of controls by distinguishing between passive and active measures and commenting that with hydrogen, the use of active measures can be a challenge because of the issues of wide flammability and high reactivity. Closest to ISD in terms of effectiveness, passive measures for facilitating hydrogen safety, including blast walls [15] and vent relief systems [16–19], have been discussed in both the archival literature and the popular media from the perspectives of success and failure of such devices.

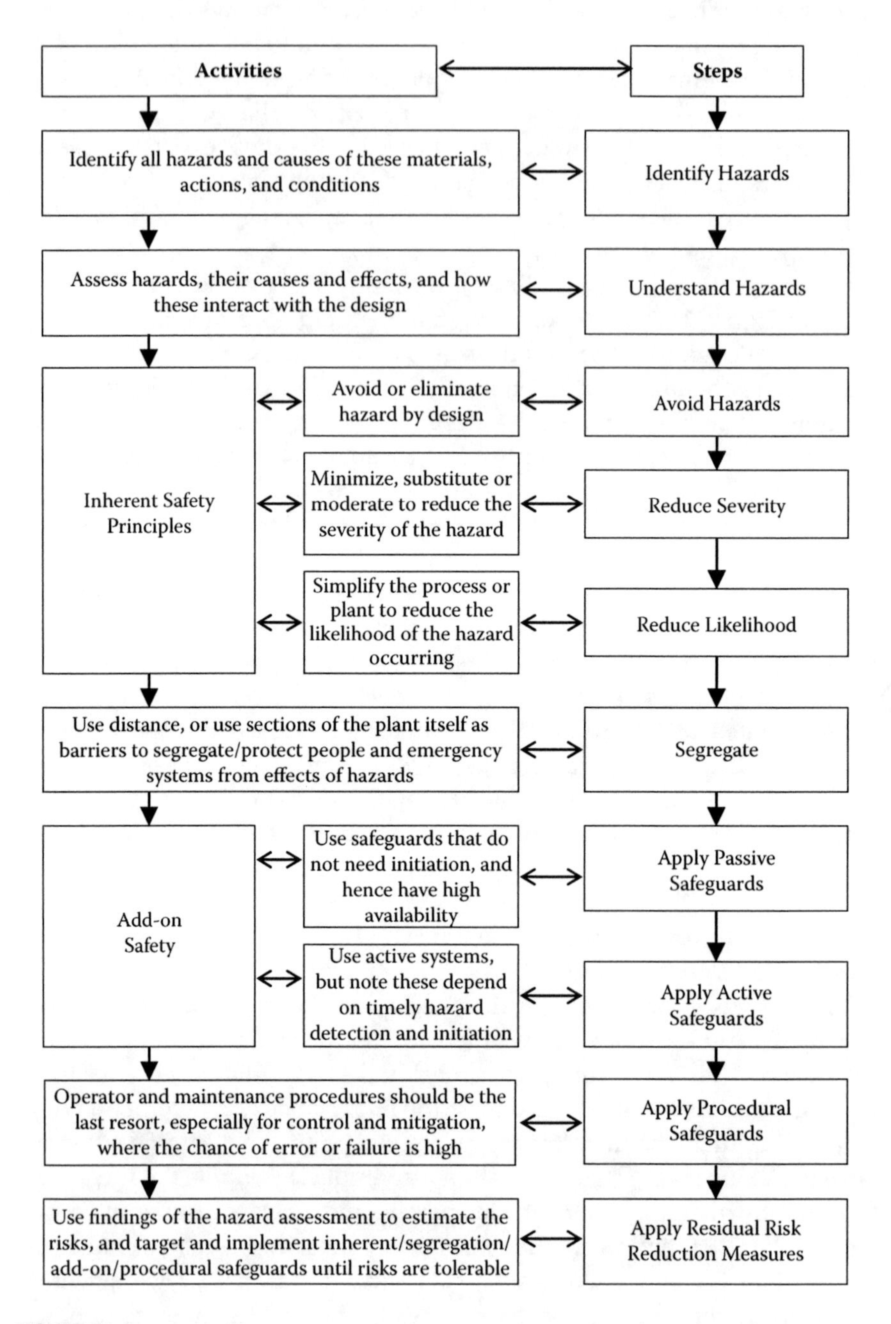

FIGURE 7.1
A systematic approach to loss prevention: the hierarchy of controls. (From Kletz, T. and Amyotte, P., *Process Plants: A Handbook for Inherently Safer Design*, 2nd edition, CRC Press/Taylor & Francis Group, Boca Raton, FL, 2010. With permission.)

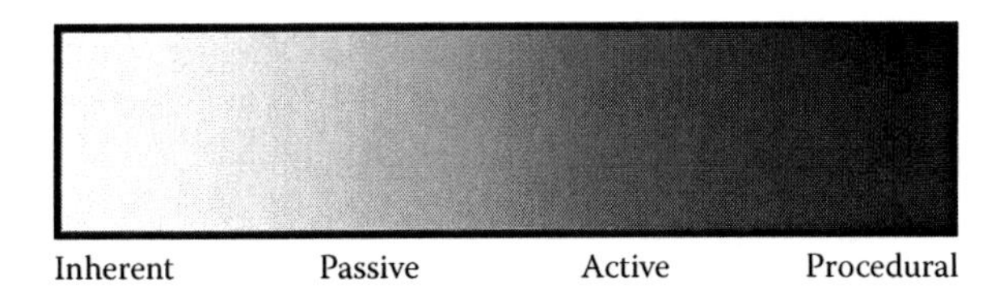

FIGURE 7.2
Representation of safety strategies as a spectrum of options from inherent through to procedural. (From Hendershot, D.C., *Process Safety Progress*, 29 (4), 389–392, 2010. With permission.)

7.2 Minimization (Intensification)

The need to avoid accumulation of flammable clouds of hydrogen is a key motivating factor for adoption of the minimization principle whenever possible. Reference [14] recommends minimizing confinement to achieve this goal and comments that "the optimal wall is no wall." This is not a universal recommendation, however, as large-momentum horizontal leaks might be better directed upwards (thus putting to good use the strong positive buoyancy of hydrogen) by the placement of vertical walls around possible leak locations [14]. Minimizing congestion in a process area is also highly beneficial with respect to limiting turbulent flame acceleration and destructive overpressure generation should ignition of a hydrogen cloud occur [14]. Pasman and Rogers [12], on the issue of minimizing release quantities, note the stringent requirement for leak-free connection technologies when handling hydrogen.

In their review of infrastructure options for hydrogen refueling stations, Markert et al. [7] comment that current scenarios are still in the very early stages of development and therefore offer cost-saving opportunities as would be brought about by the ISD approach. This advice acknowledges the fact that inherent safety is typically most effective when considered early in the design sequence. While it is possible to retrofit ISD principles to an existing plant to some extent, incorporation of inherent safety thinking in preliminary hazard and risk assessments can be highly beneficial. This is particularly important with respect to minimization, that is, storing as little hydrogen as possible [7].

Markert et al. [7] present a comparison of three hydrogen refueling alternatives with the current system for gasoline (petrol) production, storage, distribution, and dispensing, as shown in Figure 7.3. The process chain for centralized production and either truck or pipeline distribution (i.e., the first two nonpetroleum options in Figure 7.3) is shown schematically in Figure 7.4. Here one sees the need for careful consideration of the inventories to be stored both centrally and at medium-scale. The on-site option, on the other hand, eliminates the need for medium-scale storage—a case of 100 percent minimization. There are, of course, a myriad of safety, environmental,

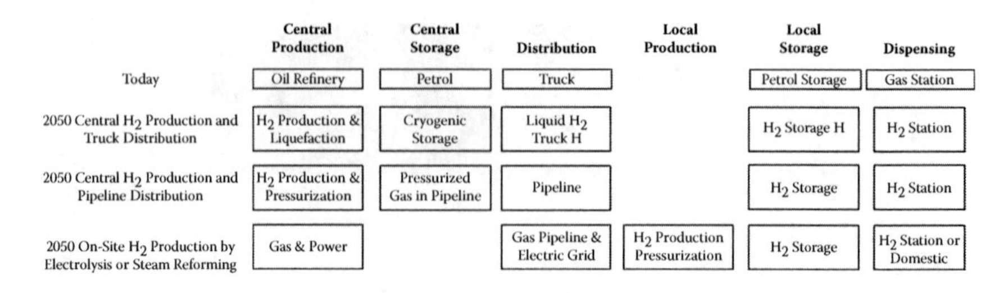

FIGURE 7.3
Comparison of gasoline and hydrogen production, storage, distribution, and dispensing schemes. (From Markert, F. et al., *International Journal of Hydrogen Energy*, 32 (13), 2227–2234, 2007. With permission.)

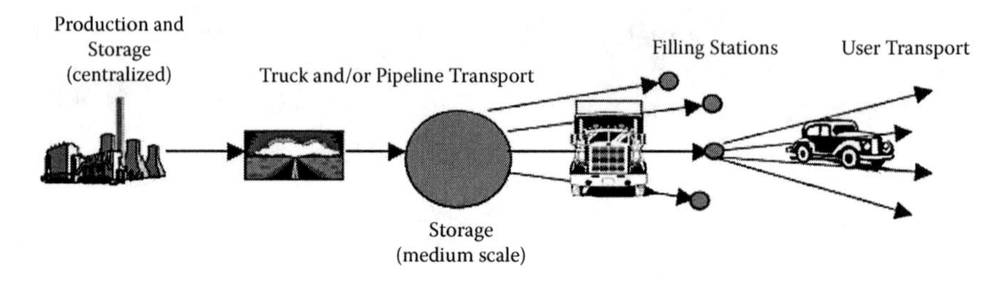

FIGURE 7.4
Process chain for centralized production of hydrogen and distribution by truck or pipeline. (From Markert, F. et al., *International Journal of Hydrogen Energy*, 32 (13), 2227–2234, 2007. With permission.)

and cost factors that must be considered in selecting from the full range of such alternatives. The point being made here is that ISD should be one of these factors.

Another interesting example of minimization is given by the experimental work of Alsheyab, Jiang, and Stanford [20], this time from the perspective of minimizing hydrogen formation during the electrochemical production of ferrate. (The term *ferrate* is often used to mean ferrate(VI), although according to IUPAC naming conventions, it may also refer to other iron-containing oxyanions, such as ferrate(V) and ferrate(IV). Ferrate(VI) refers either to the anion $[FeO_4]^{2-}$, in which iron is in the +6 oxidation state, or to a salt containing this anion.) Ferrate is described by the authors as a strong oxidant used for organic synthesis and water/wastewater treatment, and having several advantages over other commonly used oxidants such as chlorine, hydrogen peroxide, and ozone [20]. They further describe the electrochemical production of ferrate as being preferable to chemical production because of the simplicity of operation, improved safety features, and avoidance of the toxic compound hypochlorite [20]. Using the experimental set-up for lab-scale ferrate production shown schematically in Figure 7.5, Alsheyab, Jiang, and Stanford [20] determined that the hydrogen concentrations formed at the cathode were significantly less than the

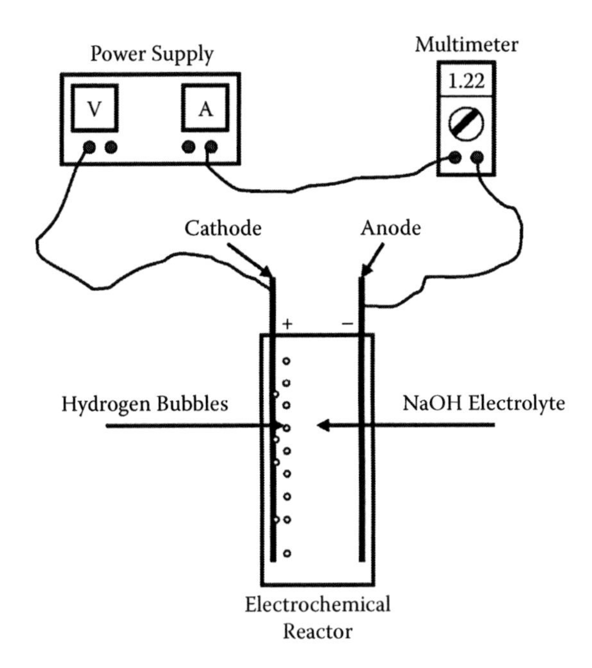

FIGURE 7.5
Ferrate production at the laboratory scale. (From Alsheyab, M., Jiang, J.-Q., and Stanford, C., *Journal of Chemical Health & Safety*, 15 (5), 16–20, 2008. With permission.)

lower flammability limit for hydrogen in air. Whether this ISD feature brought about by process intensification would hold at the pilot and production plant scales was recognized as a key risk factor requiring further investigation [20].

7.3 Substitution

As previously described, one interpretation of the substitution principle is the replacement of a substance with a less hazardous material. This clearly demonstrates the fact that inherent safety must be viewed as hazard specific. For example, while the substitution of nitrogen for natural gas as a purge gas for cleaning pipelines does eliminate the flammability hazard associated with natural gas, it also introduces the asphyxiation hazard that accompanies the use of nitrogen [21]. Thus, if one were to consider substituting some other material for hydrogen in a given application, the hazard being avoided must be specified (as in the familiar substitution of nonflammable helium for flammable hydrogen when a lighter-than-air gas is desired).

An extension of the point made in the previous sentence is the use of other substances not as a substitute for hydrogen, but as a means of avoiding hydrogen generation. Pebble-bed nuclear reactors utilizing helium rather than water as a coolant offer the inherently safer advantage of eliminating water from the reactor core. The release of hydrogen gas in the event of an upset is therefore also eliminated [22].

It is, however, the features of hydrogen associated with it being an abundant, cleaner-burning fuel that make it attractive as a substitute for other fuels such as hydrocarbons. It is necessary, therefore, to also invoke the second interpretation of substitution—that of replacing a processing route with one that does not involve hazardous material. A relevant example here is the production of the insecticide carbaryl by two different synthesis routes utilizing the same raw materials but reacting them in a different order [3]. One of the processes produces methyl isocyanate (MIC) as a hazardous intermediate while the other avoids the production of this chemical that is forever linked with the 1984 tragedy in Bhopal, India (see Section 10.2).

The example in the previous section of minimization of hydrogen inventories in storage also fits with the current discussion on substituting one process route for another. Although it was shown earlier that hydrogen can be transported by pipeline or truck from centralized plants [7], a more practical, cost-effective production method is on-site catalytic reforming from natural gas [23, 24]. While this latter approach does address the issue of large inventories of hydrogen, it also has a significant environmental impact because of the carbon dioxide generated during the process [24]. To ease this and other problems, Guy [24] comments on the importance of indirect and direct production of hydrogen by renewable energy sources and by sunlight, respectively. These are also examples of synthesis route substitution.

7.4 Moderation (Attenuation)

Pasman and Rogers [12] note that hydrogen can be shipped and stored in various forms: (1) as a compressed gas, (2) absorbed on a substrate material (e.g., as a metal hydride), and (3) as a cryogenic liquid. Each of these options is briefly considered in the following discussion from an inherent safety (moderation) perspective. While not always able to eliminate a given hazard, moderating or attenuating the form of a material or the conditions under which it is processed can be beneficial in terms of risk reduction.

Storing hydrogen as a liquid is attractive because of the high energy density per unit volume [25] (or storage volume efficiency [12]). Potential downsides include the need for heavy, bulky cryogenic tanks [26] (particularly in the case of hydrogen as a transportation fuel), as well as the intrinsic hazards of very low temperatures given that the normal boiling point of hydrogen is –252.9°C (see Chapter 3). Cryogenic temperatures also render traditional detection methods such as marking dyes, radioactive tracing gases, and odorants ineffective [27].

Hydrogen as a gas, of course, presents a significant flammability hazard. Applying the ISD principle of moderation by lowering the gas temperature, moderate reductions in the lower flammability limit of hydrogen can be achieved; the lower limit at hydrogen's normal boiling point (–252.7°C) is 7.8 percent by volume in air compared with 4 percent by volume in air at 25°C and atmospheric pressure [28]. Lowering the temperature also brings about a decrease in the upper flammability limit and therefore a narrowing of the flammability range. Varying the gas concentration away from stoichiometric conditions can help to moderate the flammability hazard. The burning velocity of hydrogen is 3.25 m/s at just above its stoichiometric concentration (29.5 percent by volume in air) at 25°C and atmospheric pressure; the value at the lower flammability limit at these conditions is only 0.04 m/s [28].

Rainer [28] notes that as with other combustible gases, the flammability range of hydrogen can be reduced by the addition of inert gases such as carbon dioxide or nitrogen. Hendershot's point that the hierarchy of controls is in fact a spectrum of options often having indistinct or blurred boundaries [5] is apparent here, especially if the nature of the hazardous material (i.e., hydrogen) is moderated by the addition of nonhazardous materials (i.e., CO_2 or N_2, at least from the perspective of flammability) with the aid of mechanical devices such as sensors and alarms. The question also remains as to the feasibility of using a diluted hydrogen stream when it is the flammability properties of hydrogen that have made it the fuel of choice in a given application. These considerations indicate again the need for trade-offs when attempting to reduce risk via a prioritized ranking of safety measures and also achieve desired operational goals.

A recent article by Middha, Engel, and Hansen [29] addresses the issue of maintaining fuel viability while minimizing hazards and the ensuing risk. The objective of this work was to investigate the addition of hydrogen to natural gas (methane) to create a hybrid fuel, hythane, having a reduced explosion risk compared with the pure component fuels. Because the proposed percentages of hydrogen were less than 50 percent, this could, strictly speaking, be viewed as moderating the explosion risk of methane as opposed to moderating the similar risk for hydrogen. Additionally, because hythane is a different fuel, this could also be viewed as an example of substitution—thus demonstrating the complementary nature of the various principles of ISD.

The motivation for the work of Middha, Engel, and Hansen [29] is reported to be the need for a stopgap fuel between fossil fuels and hydrogen. Fuel blends of 8 to 30 percent hydrogen in methane offer the potential for reduced emissions with no significant modifications to existing infrastructure [29]. According to the computational fluid dynamics results obtained in this work, the combined risk of dispersion and explosion for hythane was determined to be comparable to that of methane (or lower in some instances) and lower than that of hydrogen in all cases studied [29].

Metal hydrides provide a means to store hydrogen in solid form [25], and offer an inherently safer approach in some regards to liquid and gaseous storage [12]. Pasman and Rogers [12] further comment, however, that while the hydride itself can be viewed as inherently safer, production of hydrides with acceptable risk and retrieval of the hydrogen at moderate temperatures remain challenging and hence the subject of intensive research.

Yang [30] states that metal hydrides (as well as chemical, carbon-based, and advanced-material hydrides) can potentially store large quantities of hydrogen at low pressures. Low-pressure storage is a classic application of the moderation principle; large inventories are essentially a negation of the minimization principle. The issue of trade-offs appears once more; risk assessments must therefore consider the exposure of hydride systems and their potentially large inventories of hydrogen to high temperatures and the resulting pressure increase brought about by flame impingement from an accidental fire [30].

7.5 Simplification

The ISD principle of simplification is widely applicable in all industrial endeavors. Simplification can be thought of not only in relation to inherent safety but also with respect to passive and active engineered safety devices as well as procedural measures. Barriers and sensors that are simple yet effective provide fewer opportunities for a malfunction to occur. Similarly, an easy-to-understand operating procedure is more likely to be followed than one that is overly complex and seemingly written by people who have never had to perform the given procedure.

One approach to simplification as an ISD measure is to make equipment robust enough to withstand any foreseeable undesired event such as a pressure excursion. This philosophy seems well suited to the hydrogen industry as a possible means of reducing the reliance on explosion relief vents. Cost is an obvious consideration in using thick-walled vessels, which some practitioners would view as a passive rather than an inherent measure.

These points on simplification were considered by Xu et al. [31] in their study of multilayered stationary high-pressure hydrogen storage vessels (SHHSVs). They give the following list of SHHSV characteristics aimed at enhancing inherent safety [31]:

- As uniform a stress distribution as possible
- As few welds as possible
- High fatigue resistance to avoid failure from pressure swings caused by repeated vessel filling and discharging

- A convenient arrangement for online leak monitoring
- Hydrogen-compatible materials of construction

The stipulation of as few welds as possible is also related to the concept of minimizing potential leak and failure locations (welds, flanges, etc.).

Simplification of an overall process can also help to avoid incidents. Guy [24] describes the Chicago Transit Authority Zero Emission Fuel Cell Bus system in which buses store compressed hydrogen in gas cylinders for use in onboard fuel cells. Each bus has a range of 250 miles; what could potentially be a complex, distributed hydrogen refueling system has been simplified by having the buses refuel at a central hub in a time of 15 minutes [24].

Janssen et al. [32] conducted research on the development of a high-pressure electrolyser with the aim of achieving higher system efficiency for hydrogen production. While conventional electrolytic hydrogen production uses a low-pressure process (an example of ISD through moderation), the process proposed by Janssen et al. [32] employed the ISD principle of simplification to eliminate the need for compressors and buffer tanks from the overall design.

7.6 Other Examples

Two ISD subprinciples provide other examples of inherent safety application to the hydrogen industry: (1) *making incorrect assembly impossible* (simplification) and (2) *limitation of effects* (moderation). As described by Kletz and Amyotte [3], it is not uncommon for the second of these subprinciples to be equated with *avoiding domino or knock-on effects*. This most often happens when limitation of the effects of a hazard is undertaken not only at the hazard source but also with respect to far-field consequences. In this manner and as shown in Figure 7.1, unit segregation provides a bridge in the continuum between ISD and passive barriers.

The U.S. Chemical Safety Board (CSB) has issued a safety bulletin [33] on positive material verification that is pertinent to the matter of taking steps to avoid incorrect assembly [1]. During shutdown of the plant in question, three elbows were removed from a high-pressure, high-temperature hydrogen line for cleaning. Because of different process conditions in the line, two of the elbows were composed of a stronger metal alloy, while the third was constructed of less expensive carbon steel. During reinstallation, the carbon steel elbow was incorrectly positioned and subsequently corroded, leading to a hydrogen release and fire that caused extensive property damage and minor injury to an employee. A key finding identified by the CSB in relation to this process incident is the following:

Piping systems can be designed such that incompatible components cannot be interchanged. All three elbows could have been made from the same low alloy steel material, even though this would have meant additional expense. Alternatively, elbow 1 could have been dimensionally different from elbow 2 and 3, although this would have meant additional construction costs [33].

One approach to limiting the near-field effects of an incident is given by Segal, Wallace, and Keffer [34], who described their fuel supply system to provide gaseous hydrogen to an engine test cell. A key feature of their design was the outdoor storage of 70.5 m^3 of hydrogen in cylinders. While engineered safety devices and presumably procedural measures were also incorporated, separation of the fuel source from operational personnel is an example of simplification through limitation of effects. It should be noted, however, that outside storage of compressed gaseous hydrogen requires consideration of phenomena such as turbulent jet flames arising from a fuel leak and subsequent ignition [12].

Avoidance of domino effects (i.e., damage to adjacent units and harm to people occupying them) is a critical aspect of hydrogen safety [13, 14]. The topic of safety distances (or harm thresholds) with respect to hydrogen dispersion, deflagration, and detonation is therefore an area of intensive research (e.g., [35–37]). A representative example from this body of work is given in Figure 7.6, which shows a nomograph of blast radiuses corresponding to different damage levels for various masses of hydrogen released and ignited. Figure 7.6 comes from the modeling work of Dorofeev [36] and represents the case of low congestion such as might be experienced within a hydrogen refueling station.

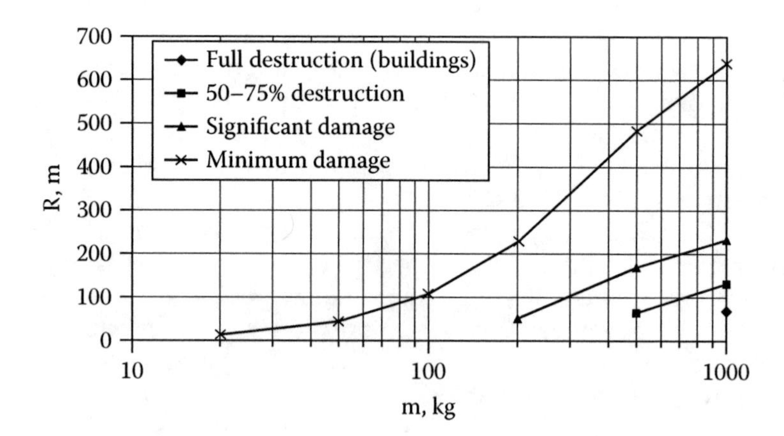

FIGURE 7.6
Blast radii for various degrees of building damage as a function of mass of hydrogen released in the case of low congestion. (From Dorofeev, S.B. *International Journal of Hydrogen Energy*, 32 (13), 2118–2124, 2007. With permission.)

7.7 Measurement of Inherent Safety

As the previous sections have demonstrated, the principles of inherently safer design are widely applicable to the production, distribution, and utilization of hydrogen for a variety of purposes. This section addresses the issue of how to decide on the most appropriate application of these principles drawn from a range of alternatives.

In general, a relatively large number of ISD assessment tools exist for attempting to measure the degree of inherent safety of a given process or processing methodology. These techniques have been reviewed by, among others, Khan and Amyotte [38], Khan, Sadiq, and Amyotte [39], and Kletz and Amyotte [3].

One of the earliest of such efforts in the 1990s was the INSIDE (*IN*herent *S*HE [Safety, Health, Environment] *In DE*sign) project conducted in Europe, and which resulted in the INSET (*IN*herent *S*HE *E*valuation *T*ool) toolkit to identify inherently safer design options throughout the life of a process and to evaluate the options [3]. There are also well-known process safety methodologies such as checklist and what-if analyses that have been adapted for ISD consideration, largely in terms of identifying hazards that can be addressed by the principles of inherent safety.

Well-established indexing methods such as the Dow Fire and Explosion Index (F&EI) and Dow Chemical Exposure Index (CEI) have numerous inherent safety aspects associated with their calculation procedures. Etowa et al. [40] quantified the ISD features of both Dow indices (F&EI and CEI) and demonstrated the beneficial impact of the principles of minimization, substitution, and moderation. The work of Edwards and coworkers at Loughborough University in the United Kingdom resulted in one of the first indices designed specifically to address inherently safer processing opportunities, the prototype index of inherent safety, or PIIS [41, 42].

Over the past decade or so there has been a proliferation of ISD indexing procedures and assessment methodologies appearing in the process safety literature. Table 7.2 gives a summary of such techniques developed over the period 2002–2010. Observations on the entries in Table 7.2 include the following [3]:

- Many of the methods deal specifically with the early concept and route-selection stages of the design process.
- Some of the approaches use sophisticated mathematical and problem-solving techniques such as fuzzy logic.
- There has been a growing trend to link inherent safety with environmental and health issues in an effort to achieve an integrated approach.
- There have been attempts to incorporate inherent safety assessment into process design simulators.

TABLE 7.2

Examples of Development of Inherent Safety Assessment Methodologies and Elucidation of Assessment Considerations (2002–2010)

Reference	Contribution
Khan and Amyotte [43]	Integrated Inherent Safety Index (I2SI).
Khan and Amyotte [44]	Further development of I2SI to include cost model.
Carvalho et al. [45]	Method for identifying retrofit design alternatives of chemical processes. Uses Inherent Safety Index (ISI) developed by other researchers.
Hurme and Rahman [46]	Discussion of implementation of inherent safety throughout process life cycle phases. Use of Inherent Safety Index (ISI) developed earlier.
Rahman et al. [47]	Comparative evaluation of three inherent safety indices with expert judgment at process concept phase.
Hassim and Hurme [48]	Inherent Occupational Health Index developed to assess the health risk of process routes during the process research and development stage. The index can be used to compare process routes or to determine the level of inherent occupational health hazards.
Hassim and Hurme [49]	Occupational Health Index (OHI) developed for assessment during the basic engineering stage. This method relies on the information available in piping and instrumentation diagrams and the plot plan. The health aspects considered are chronic and acute inhalation risks, and dermal/eye risk.
Hassim and Hurme [50]	Health Quotient Index (HQI) developed for assessment during the preliminary process design phase. This index quantifies a worker's health risk from exposure to fugitive emissions by using data from process flow diagrams. This method can be used to quantify the level of risk from a process or to compare alternative processes.
Hassim and Hurme [51]	Method to estimate inhalation exposures and risks; for use early in the design stages by utilizing process flow diagrams. The risk of chemical exposure can be evaluated either through the hazard quotient method or by calculating the carcinogenic intake and resulting cancer risk.
Hassim and Edwards [52]	Process Route Healthiness Index (PRHI) for quantification of health hazards arising from alternative chemical process routes. Application is in early stages of chemical plant design.
Gupta and Edwards [53]	Graphical approach for evaluating inherent safety based on earlier developed Loughborough Prototype Index of Inherent Safety (PIIS).
Cozzani et al. [54]	Procedure for assessment of hazards arising from decomposition products formed due to loss of chemical process control. Applicable to consideration of substitution principle.
Landucci et al. [55]	Procedure and indices for evaluating inherent safety at preliminary process flow diagram (PFD) stage for hydrogen storage options.
Landucci et al. [56]	Consequence-based method for identification and assessment of inherently safer plant design alternatives in developing steam reforming processes for hydrogen production.
Landucci et al. [57]	Further development of PFD method [55] by use of quantitative key performance indicators (KPIs) to remove subjective judgment.

TABLE 7.2 (*Continued*)

Examples of Development of Inherent Safety Assessment Methodologies and Elucidation of Assessment Considerations (2002–2010)

Reference	Contribution
Tugnoli et al. [58]	Quantitative inherent safety assessment method utilizing process flow diagrams (PFDs) in early design stages for hydrogen supply chain options. The result of the assessment is a quantification of the inherent safety of the process scheme by a set of key performance indicators (KPIs).
Landucci et al. [59]	Consequence-based approach for inherent safety assessment of production, distribution, and utilization systems with respect to hydrogen vehicles.
Cordella et al. [60]	Further development of procedure for decomposition product analysis [54] to account for acute and long-term harm to human health, ecosystem damage, and environmental media contamination.
Shariff et al. [61]	Integrated Risk Estimation Tool (iRET) for inherent safety application at preliminary design stage. iRET links the design simulation software HYSYS with an explosion consequence model.
Leong and Shariff [62]	Further development of iRET [61] to incorporate a quantitative Inherent Safety Level (ISL), thus enabling integration of design simulation software with an Inherent Safety Index Module (ISIM). Application is again at the preliminary design stage.
Leong and Shariff [63]	Evolution of ISIM [62] to a Process Route Index (PRI) for comparison and ranking of different routes to manufacture the same product based on hazard potential of routes.
Shariff and Leong [64]	Method of evaluating inherent risk within a process as a result of the chemicals used and the process conditions. Through integration with HYSYS, the method can be used as early as the initial design stages to determine the probability and consequence of possible risk due to major accidents.
Shariff and Zaini [65]	Development of toxic release consequence analysis tool (TORCAT), a tool for consequence analysis and design improvement through use of inherent safety principles. The method utilizes an integrated process design simulator and a toxic release consequence analysis model.
Rusli and Shariff [66]	Qualitative assessment for inherently safer design (QAISD) method for application during preliminary design. This qualitative method combines hazard review techniques with inherently safer design concepts to generate inherently safer plant options/proactive measures.
Kossoy and Akhmetshin [67]	Use of nonlinear optimization method to select inherently safer operational parameters for given configuration of reactor equipment and materials. Primary concern is cooling failure.
Shah et al. [68]	SREST (substance, reactivity, equipment and safety technology) layer assessment method for environment, health and safety (EHS) aspects in early phases of chemical process design.
Adu et al. [69]	Comparative evaluation of various methods for assessing EHS hazards in early phases of chemical process design.
Palaniappan et al. [70]	Methodology for integrated inherent safety and waste minimization analysis during process design.

—continued

TABLE 7.2 (*Continued*)

Examples of Development of Inherent Safety Assessment Methodologies and Elucidation of Assessment Considerations (2002–2010)

Reference	Contribution
Palaniappan et al. [71]	Indexing procedure for inherent safety analysis at process route selection stage.
Palaniappan et al. [72]	Indexing procedure for inherent safety analysis at process flowsheet development stage. Discussion of *i*Safe, an expert system for automating procedures developed by Palaniappan et al. [71, 72].
Srinivasan and Nhan [73]	Inherent benignness indicator (IBI), a statistical analysis-based method for comparing alternative chemical process routes.
Srinivasan and Kraslawski [74]	Application of TRIZ methodology for creative problem solving to design of inherently safer chemical processes.
Meel and Seider [75]	Use of game theory to achieve inherently safer operation of chemical reactors.
Gentile et al. [76]	Fuzzy logic-based index for evaluation of inherently safer process alternatives with the aim of linking to process simulation.
Al-Mutairi et al. [77]	Linking of inherent safety and environmental concerns with optimization of process scheduling.

Source: Data for 2002–2008 from Kletz and Amyotte [3].

- Some of the indexing methods have been in existence long enough for comparative evaluations to be made among them.

When commenting in 2005 on potential barriers to wider adoption of inherently safer design principles in the process industries, Edwards [78] noted that the issue may not be the availability of inherent safety assessment tools but rather the limited use of these tools by industry. Reasons might include the subjective judgment required by some of these tools and also their attendant complexity [3]. In 2011, it seems that the same availability of tools, yet limited uptake by industry in general, exists. It is too early to comment in this regard with specific reference to the hydrogen industry, as hydrogen-specific ISD assessment methodologies have only recently appeared in the process safety literature (2007–2010).

While some of the more general entries in Table 7.2 could undoubtedly be applied to ISD assessments involving hydrogen, it does appear that the most concerted effort concerning hydrogen ISD evaluation has come from the team with members based in Pisa and Bologna, Italy [55–59]. This series of publications provides an interesting and valuable contribution to the field of hydrogen safety. Beginning in the early design stages and with the basic method shown in Figure 7.7, ISD assessments of storage [55] and production [56] alternatives were extended to include distribution and utilization considerations generally [58] and specifically with respect to the automotive sector [59]. Key performance indicators (KPIs) were introduced to account for unit, overall, and domino hazard indices [57].

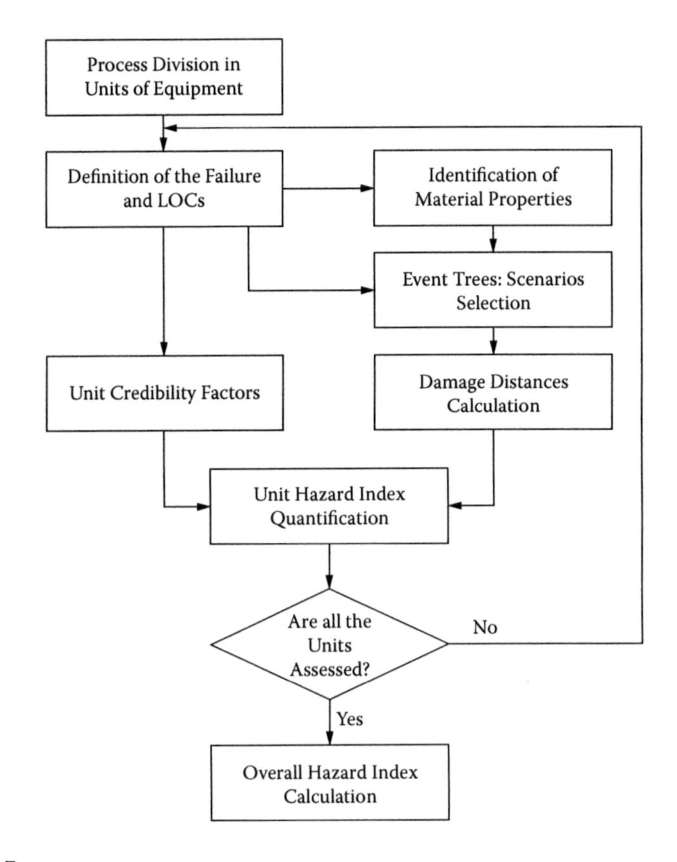

FIGURE 7.7
Flow chart of hydrogen-specific ISD assessment method, where LOC stands for loss of containment. (From Landucci, G. et al., Assessment of inherently safer technologies for hydrogen production, in Proceedings of the 5th International Seminar on Fire and Explosion Hazards, Edinburgh, UK, April 23–27, 2007. With permission.)

Several features of this body of work [55–59] are noteworthy in terms of potential adoption of the developed methodologies by industrial designers and other practitioners. The first is the previously mentioned emphasis on early design stages when ISD considerations typically have the greatest impact. Second, there is a clear recognition that the preferred term is inherently *safer* design, not inherently *safe* design. This is demonstrated in Figure 7.8 which facilitates, for example, comparisons between cryogenic storage of liquefied hydrogen (schemes *a* and *b*) and bulk storage of metal hydrides (scheme *c*). A third important feature of this work is the explicit and unambiguous use of inherent safety terminology, for example, *substitution* and *moderation* when examining metal hydride options [58], as well as a clear placement of ISD technologies within the hierarchy of controls. All of these points are key to the advancement of the inherent safety concept within the various sectors of the hydrogen industry.

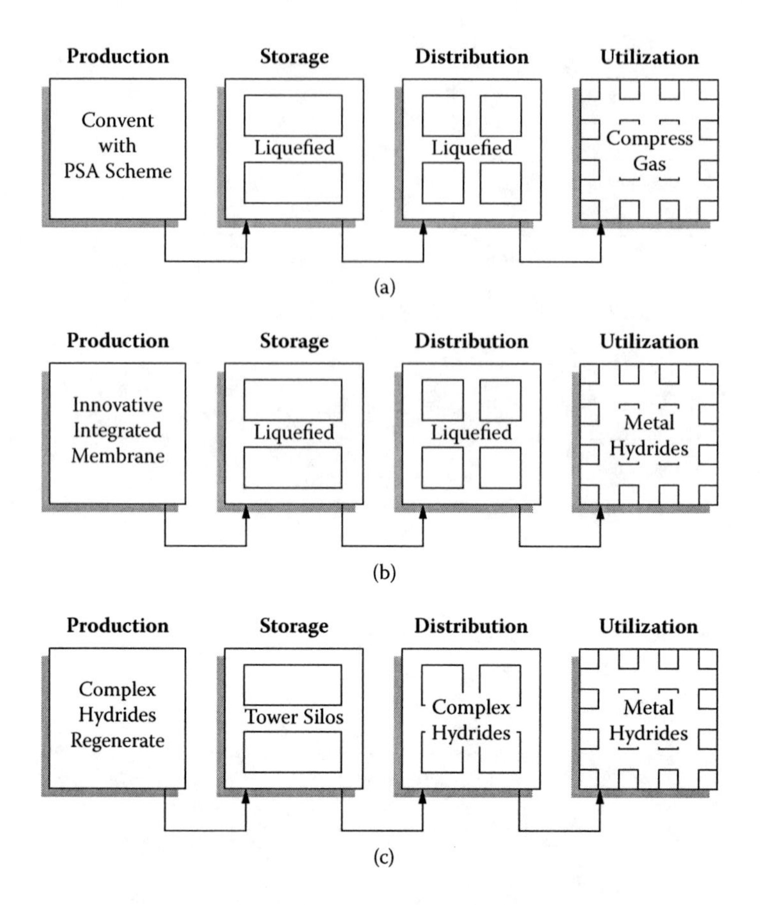

FIGURE 7.8
Alternative hydrogen supply chains for ISD assessment. (From Tugnoli, A., Landucci, G., and Cozzani, V., *Process Safety Progress*, 28 (2), 156–170, 2009. With permission.)

References

1. Amyotte, P.R., MacDonald, D.K., and Khan, F.I., An analysis of CSB investigation reports for inherent safety learnings, Paper No. 44a, *Proceedings of 13th Process Plant Safety Symposium*, 7th Global Congress on Process Safety (AIChE 2011 Spring National Meeting), Chicago, IL (March 13–16, 2011).
2. CCPS, *Inherently Safer Chemical Processes. A Life Cycle Approach*, 2nd edition, John Wiley & Sons, Hoboken, NJ, 2009.
3. Kletz, T., and Amyotte, P., *Process Plants: A Handbook for Inherently Safer Design*, 2nd edition, CRC Press/Taylor & Francis Group, Boca Raton, FL, 2010.
4. Khan, F.I., and Amyotte, P.R., How to make inherent safety practice a reality, *Canadian Journal of Chemical Engineering*, 81 (1), 2–16, 2003.
5. Hendershot, D.C., A summary of inherently safer technology, *Process Safety Progress*, 29 (4), 389–392, 2010.

6. Crowl, D.A., and Jo, Y.-D., The hazards and risks of hydrogen, *Journal of Loss Prevention in the Process Industries*, 20 (2), 158–164, 2007.
7. Markert, F., Nielsen, S.K., Paulsen, J.L., and Andersen, V., Safety aspects of future infrastructure scenarios with hydrogen refuelling stations, *International Journal of Hydrogen Energy*, 32 (13), 2227–2234, 2007.
8. General Hydrogen Corporation, *Material Safety Data Sheet: Compressed Hydrogen*, Washington, PA (undated).
9. Hydrogen Association, *Hydrogen Safety*, Fact Sheet Series (undated).
10. Molkov, V., *Hydrogen Safety Research: State-of-the-Art*, Proceedings of the 5th International Seminar on Fire and Explosion Hazards, Edinburgh, UK (April 23–27, 2007).
11. Miller, M., *Hydrogen Fueling Stations*, Institute of Transportation Studies (November 15, 2004).
12. Pasman, H.J., and Rogers, W.J., Safety challenges in view of the upcoming hydrogen economy: An overview, *Journal of Loss Prevention in the Process Industries*, 23 (6), 697–704, 2010.
13. Roads2HyCom, *Hydrogen Safety Measures*, Document Tracking ID 5031, www.roads2hy.com (March 30, 2011).
14. HySafe, *Chapter V: Hydrogen Safety Barriers and Safety Measures*, Biennial Report on Hydrogen Safety, Version 1.0 (May 2006).
15. Groethe, M., Merilo, E., Colton, J., Chiba, S., Sato, Y., and Iwabuchi, H., Large-scale hydrogen deflagrations and detonations, *International Journal of Hydrogen Energy*, 32 (13), 2125–2133, 2007.
16. Benard, P., Mustafa, V., and Hay, D.R., Safety assessment of hydrogen disposal on vents and flare stacks at high flow rates, *International Journal of Hydrogen Energy*, 24 (5), 489–495, 1999.
17. Astbury, G.R., Venting of low pressure hydrogen gas: A critique of the literature, *Process Safety and Environmental Protection*, 85 (4), 289–304, 2007.
18. Wald, M.L., and Pollack, A., Core of stricken reactor probably leaked, U.S. says, *New York Times* (April 6, 2011).
19. Tabuchi, H., Bradsher, K., and Wald, M.L., In Japan reactor failings, danger signs for the U.S., *New York Times* (May 17, 2011).
20. Alsheyab, M., Jiang, J.-Q., and Stanford, C., Risk assessment of hydrogen gas production in the laboratory scale electrochemical generation of ferrate (VI), *Journal of Chemical Health & Safety*, 15 (5), 16–20, 2008.
21. CSB, *Urgent Recommendations from Kleen Energy Investigation*, U.S. Chemical Safety and Hazard Investigation Board, Washington, D.C., 2010.
22. Bradsher, K., Pressing ahead where others have failed, *New York Times* (March 24, 2011).
23. Sherman, D., At milepost 1 on the hydrogen highway, *New York Times* (April 29, 2007).
24. Guy, K.W., The hydrogen economy, *Process Safety and Environmental Protection*, 78 (4), 324–327, 2000.
25. Motavalli, J., A universe of promise (and a tankful of caveats), *New York Times* (April 29, 2007).
26. Leary, W.E., Use of hydrogen as fuel is moving closer to reality, *New York Times* (April 16, 1995).
27. Leary, W.E., With shuttle back in space, NASA returns to leak problem, *New York Times* (October 9, 1990).

28. Rainer, D., Hydrogen, *Journal of Chemical Health and Safety*, 15 (4), 49–50, 2008.
29. Middha, P., Engel, D., and Hansen, O.R., Can the addition of hydrogen to natural gas reduce the explosion risk? *International Journal of Hydrogen Energy*, 36 (3), 2628–2636, 2011.
30. Yang, J.C., Material-based hydrogen storage, *International Journal of Hydrogen Energy*, 33 (16), 4424–4426, 2008.
31. Xu, P., Zheng, J., Liu, P., Chen, R., Kai, F., and Li, L., Risk identification and control of stationary high-pressure hydrogen storage vessels, *Journal of Loss Prevention in the Process Industries*, 22 (6), 950–953, 2009.
32. Janssen, H., Bringmann, J.C., Emonts, B., and Schroeder, V., Safety-related studies on hydrogen production in high-pressure electrolysers, *International Journal of Hydrogen Energy*, 29 (7), 759–770, 2004.
33. CSB, *Positive Material Verification: Prevent Errors During Alloy Steel Systems Maintenance*, Safety Bulletin, No. 2005-04-B, U.S. Chemical Safety and Hazard Investigation Board, Washington, D.C., 2006.
34. Segal, L., Wallace, J.S., and Keffer, J.F., Safety considerations in the design of a gaseous hydrogen fuel supply for engine testing, *International Journal of Hydrogen Energy*, 11 (11), 737–743, 1986.
35. Matthijsen, A.J.C.M., and Kooi, E.S., Safety distances for hydrogen filling stations, *Fuel Cells Bulletin*, 2006 (11), 12–16, 2006.
36. Dorofeev, S.B., Evaluation of safety distances related to unconfined hydrogen explosions, *International Journal of Hydrogen Energy*, 32 (13), 2118–2124, 2007.
37. Marangon, A., Carcassi, M., Engebo, A., and Nilsen, S., Safety distances: Definition and values, *International Journal of Hydrogen Energy*, 32 (13), 2192–2197, 2007.
38. Khan, F.I., and Amyotte, P.R., How to make inherent safety practice a reality, *Canadian Journal of Chemical Engineering*, 81 (1), 2–16, 2003.
39. Khan, F.I., Sadiq, R., and Amyotte, P.R., Evaluation of available indices for inherently safer design options, *Process Safety Progress*, 22 (2), 83–97, 2003.
40. Etowa, C.B., Amyotte, P.R., Pegg, M.J., and Khan, F.I., Quantification of inherent safety aspects of the Dow indices, *Journal of Loss Prevention in the Process Industries*, 15 (6), 477–487, 2002.
41. Edwards, D.W., and Lawrence, D., Assessing the inherent safety of chemical process routes, *Process Safety and Environmental Protection*, 71 (B4), 252–258, 1993.
42. Edwards, D.W., Rushton, A.G., and Lawrence, D., Quantifying the inherent safety of chemical process routes, Paper presented at the 5th World Congress of Chemical Engineering, San Diego, CA (July 14-18, 1996).
43. Khan, F.I., and Amyotte, P.R., Integrated Inherent Safety Index (I2SI): A tool for inherent safety evaluation, *Process Safety Progress*, 23 (2), 136–148, 2004.
44. Khan, F.I., and Amyotte, P.R., I2SI: A comprehensive quantitative tool for inherent safety and cost evaluation, *Journal of Loss Prevention in the Process Industries*, 18 (4–6), 310–326, 2005.
45. Carvalho, A., Gani, R., and Matos, H., Design of sustainable chemical processes: Systematic retrofit analysis generation and evaluation of alternatives, *Process Safety and Environmental Protection*, 86 (5), 328–346, 2008.
46. Hurme, M., and Rahman, M., Implementing inherent safety throughout process lifecycle, *Journal of Loss Prevention in the Process Industries*, 18 (4–6), 238–244, 2005.

47. Rahman, M., Heikkila, A.-M., and Hurme, M., Comparison of inherent safety indices in process concept evaluation, *Journal of Loss Prevention in the Process Industries*, 18 (4–6), 327–334, 2005.
48. Hassim, M.H., and Hurme, M., Inherent occupational health assessment during process research and development stage, *Journal of Loss Prevention in the Process Industries*, 23 (1), 127–138, 2010.
49. Hassim, M.H., and Hurme, M., Inherent occupational health assessment during basic engineering stage, *Journal of Loss Prevention in the Process Industries*, 23 (2), 260–268, 2010.
50. Hassim, M.H., and Hurme, M., Inherent occupational health assessment during preliminary design stage, *Journal of Loss Prevention in the Process Industries*, 23 (3), 476–482, 2010.
51. Hassim, M.H., and Hurme, M., Occupational chemical exposure and risk estimation in process development and design, *Process Safety and Environmental Protection*, 88 (4), 225–235, 2010.
52. Hassim, M.H., and Edwards, D.W., Development of a methodology for assessing inherent occupational health hazards, *Process Safety and Environmental Protection*, 84 (5), 378–390, 2006.
53. Gupta, J.P., and Edwards, D.W., A simple graphical method for measuring inherent safety, *Journal of Hazardous Materials*, 104 (1–3), 15–30, 2003.
54. Cozzani, V., Barontini, F., and Zanelli, S., Assessing the inherent safety of substances: Precursors of hazardous products in the loss of control of chemical systems, in Proceedings of American Institute of Chemical Engineers Spring National Meeting (2006).
55. Landucci, G., Tugnoli, A., Nicolella, C., and Cozzani, V., *Assessment of Inherently Safer Technologies for Hydrogen Storage*, IChemE Symposium Series No. 153, in 12th International Symposium on Loss Prevention and Safety Promotion in the Process Industries, Edinburgh, UK, 2007.
56. Landucci, G., Tugnoli, A., Nicolella, C., and Cozzani, V., Assessment of inherently safer technologies for hydrogen production, in Proceedings of the 5th International Seminar on Fire and Explosion Hazards, Edinburgh, UK (April 23–27, 2007).
57. Landucci, G., Tugnoli, A., and Cozzani, V., Inherent safety key performance indicators for hydrogen storage systems, *Journal of Hazardous Materials*, 159 (2–3), 554–566, 2008.
58. Tugnoli, A., Landucci, G., and Cozzani, V., Key performance indicators for inherent safety: Application to the hydrogen supply chain, *Process Safety Progress*, 28 (2), 156–170, 2009.
59. Landucci, G., Tugnoli, A., and Cozzani, V., Safety assessment of envisaged systems for automotive hydrogen supply and utilization, *International Journal of Hydrogen Energy*, 35 (3), 1493–1505, 2010.
60. Cordella, M., Tugnoli, A., Barontini, F., Spadoni, G., and Cozzani, V., Inherent safety of substances: Identification of accidental scenarios due to decomposition products, *Journal of Loss Prevention in the Process Industries*, 22 (4), 455–462, 2009.
61. Shariff, A.M., Rusli, R., Leong, C.T., Radhakrishnan, V.R., and Buang, A., Inherent safety tool for explosion consequences study, *Journal of Loss Prevention in the Process Industries*, 19 (5), 409–418, 2006.

62. Leong, C.T., and Shariff, A.M., Inherent Safety Index Module (ISIM) to assess inherent safety level during preliminary design stage, *Process Safety and Environmental Protection*, 86 (2), 113–119, 2008.
63. Leong, C.T., and Shariff, A.M., Process Route Index (PRI) to assess level of explosiveness for inherent safety quantification, *Journal of Loss Prevention in the Process Industries*, 22 (2), 216–221, 2009.
64. Shariff, A.M., and Leong, C.T., Inherent risk assessment: A new concept to evaluate risk in preliminary design stage, *Process Safety and Environmental Protection*, 87 (6), 371–376, 2009.
65. Shariff, A.M., and Zaini, D., Toxic release consequence analysis tool (TORCAT) for inherently safer design plant, *Journal of Hazardous Materials*, 182 (1–3), 394–402, 2010.
66. Rusli, R., and Shariff, A.M., Qualitative assessment for inherently safer design (QAISD) at preliminary design stage, *Journal of Loss Prevention in the Process Industries*, 23 (1), 157–165, 2010.
67. Kossoy, A., and Akhmetshin, Yu., *Simulation-Based Approach to Design of Inherently Safer Processes*, IChemE Symposium Series No. 153, in 12th International Symposium on Loss Prevention and Safety Promotion in the Process Industries, Edinburgh, UK, 2007.
68. Shah, S., Fischer, U., and Hungerbuhler, K., A hierarchical approach for the evaluation of chemical process aspects from the perspective of inherent safety, *Process Safety and Environmental Protection*, 81 (6), 430–443, 2003.
69. Adu, I.K., Sugiyama, H., Fischer, U., and Hungerbuhler, K., Comparison of methods for assessing environmental, health and safety (EHS) hazards in early phases of chemical process design, *Process Safety and Environmental Protection*, 86 (3), 77–93, 2008.
70. Palaniappan, C., Srinivasan, R., and Halim, I., A material-centric methodology for developing inherently safer environmentally benign processes, *Computers and Chemical Engineering*, 26 (4–5), 757–774, 2002.
71. Palaniappan, C., Srinivasan, R., and Tan, R., Expert system for the design of inherently safer processes, Part 1: Route selection stage, *Industrial Engineering Chemistry Research*, 41 (26), 6698–6710, 2002.
72. Palaniappan, C., Srinivasan, R., and Tan, R., Expert system for the design of inherently safer processes, Part 2: Flowsheet development stage, *Industrial Engineering Chemistry Research*, 41 (26), 6711–6722, 2002.
73. Srinivasan, R., and Nhan, N.T., A statistical approach for evaluating inherent benignness of chemical process routes in early design stages, *Process Safety and Environmental Protection*, 86 (3), 163–174, 2008.
74. Srinivasan, R., and Kraslawski, A., Application of the TRIZ creativity enhancement approach to design of inherently safer chemical processes, *Chemical Engineering and Processing*, 45 (6), 507–514, 2006.
75. Meel, A., and Seider, W.D., Dynamic risk assessment of inherently safe chemical processes: Accident precursor approach, Presented at American Institute of Chemical Engineers Spring National Meeting, Atlanta, GA, 2005.
76. Gentile, M., Rogers, W.J., and Mannan, M.S., Development of a fuzzy logic-based inherent safety index, *Process Safety and Environmental Protection*, 81 (6), 444–456, 2003.

77. Al-Mutairi, E.M., Suardin, J.A., Mannan, S.M., and El-Halwagi, M.M., An optimization approach to the integration of inherently safer design and process scheduling, *Journal of Loss Prevention in the Process Industries*, 21 (5), 543–549, 2008.
78. Edwards, D.W., Are we too risk-averse for inherent safety? An examination of current status and barriers to adoption, *Process Safety and Environmental Protection*, 83 (2), 90–100, 2005.

8

Safety Management Systems

A key engineering tool for industrial practice is a management system appropriate for the risks being addressed (process safety, occupational safety, health, environment, asset integrity, etc.) [1]. Such safety management systems are recognized and accepted worldwide as best-practice methods for managing risk. They typically consist of 10 to 20 program elements that must be effectively carried out to manage risks in an acceptable way. This need is based on the understanding that once a risk is identified, it does not go away; there is always a possibility that the adverse event will occur unless the management system is actively monitoring company operations for concerns and taking proactive actions to correct potential problems [2].

In this chapter, safety management systems are first discussed from a general perspective. This is followed by an examination of the elements that make up a typical management system for specifically dealing with issues related to process safety. The important and timely topic of safety culture is then briefly addressed. The chapter is structured in a manner similar to the discussion in Kletz and Amyotte [3] (with relevant excerpts) on the relationship between process safety management and inherently safer design; here, of course, the emphasis is on the importance of adopting a management approach to ensuring hydrogen safety.

8.1 Introduction to Safety Management Systems

As discussed in Chapter 2, root causes identified for the 1989 vapor cloud explosion in Pasadena, Texas, were a lack of both a hazard assessment study and an effective permit system for the control of maintenance activities. These factors are clearly related to the safety management system in place at the facility as they are effectively beyond the control of individual workers. Although it is incumbent upon plant operators to follow safe work procedures for maintenance and other activities, it is a management responsibility to develop those procedures in the first place and to ensure they are implemented and revised as needed. It is well established by modern incident causation theory that management system deficiencies represent the ultimate cause of industrial accidents [4].

There is thus a need for a safety management system approach in the hydrogen industry, a fact that has been recognized by various organizations. For example, the U.S. Department of Energy (DOE) has commented in its technical planning document for hydrogen safety that (with italics added here for emphasis) [5], "The continued safe operation, handling and use of hydrogen and related systems require *comprehensive safety management*."

What form, then, should such comprehensive safety management take? As noted in the introductory sentence to this chapter, safety management systems are most effective when they are tailored to the specific hazards and risks of concern. Regardless of whether the management approach is aimed at preventing occupational incidents related to individuals or process incidents that are more systemwide, however, certain features will (or should) be universal. Stelmakowich [6] describes a set of framework components for Occupational Health and Safety Assessment Series (OHSAS) 18000 that are equally applicable to other areas within the safety field:

- Continuous improvement
- Policy development and support by senior management personnel
- Planning (e.g., hazard identification, risk assessment, risk control)
- Implementation and operation (e.g., assignment of responsibilities, training, emergency preparedness and response)
- Checks and corrective actions (e.g., incident investigation, auditing)
- Management review

Management system requirements such as those given in the above list are widely known by the generic titles of *plan, do, check, act*. These four essential management functions apply to the safety management system as a whole and also to each of the individual elements comprising the management system. This point is arguably the most important feature of the management system approach to safety (hydrogen or otherwise).

The provision of additional focus for a safety management system is often addressed by reference to the familiar *incident pyramid* shown in Figure 8.1. Various studies over the years, covering a broad range of industries and typically from an occupational health and safety (OH&S) perspective, have resulted in different category totals for the pyramid levels but generally the same ratios between levels (see, for example, Bird and Germain [8]). As explained by Creedy [7]:

> The idea of the pyramid is that really serious incidents such as fatalities occur so rarely in most organizations these days that it's not practical to use them as a measure for monitoring and improving an organization's safety effectiveness—there simply aren't enough such incidents to know whether things are getting better or worse. However, it is practical

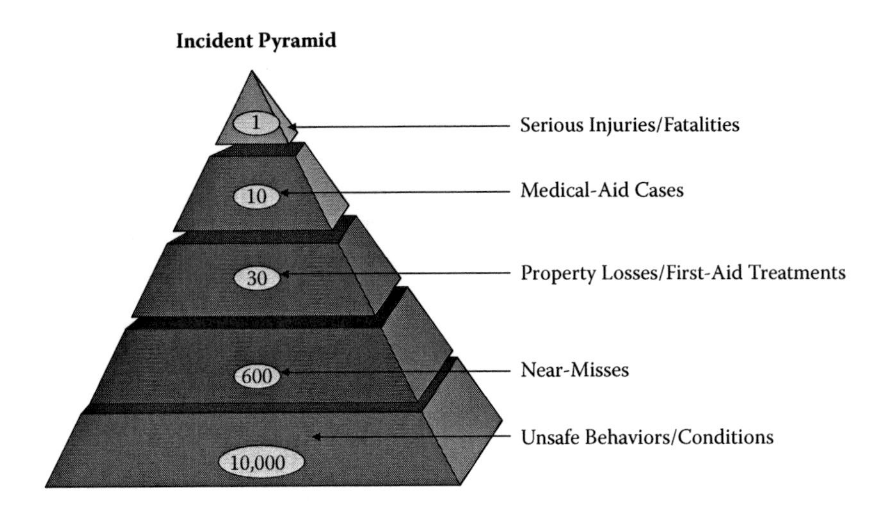

FIGURE 8.1
The incident pyramid. (From Creedy, G., *Process Safety Management*, PowerPoint presentation prepared for Process Safety Management Division, Chemical Institute of Canada, Ottawa, Ontario, 2004. With permission.)

to track the larger number of less serious incidents and use that as a performance measure.

Hence, many safety management systems—particularly those dealing with OH&S and avoidance of lost-time injuries (LTIs)—rightfully emphasize the prevention, control, and investigation of near-misses and unsafe (or at-risk) acts and conditions. Creedy [7] does, however, provide a cautionary note:

> The pyramid is certainly a useful concept. Nevertheless, it does have a serious problem that is often unrecognized, even by workplace health and safety professionals. The problem is that the conditions that can lead to really serious incidents—those which could kill or seriously injure a large number of people—may not be identified by focusing on the bottom end of the pyramid, which could thus become a distraction rather than a help.

The thinking expressed in the above passage allows us to make a further distinction between *occupational* safety and *process* safety. As previously mentioned, occupational or "traditional" safety largely aims to control individual exposures—often referred to as slips, trips, and falls. Process safety, when properly addressed with a systems approach, aims to prevent and mitigate process-related injuries and incidents. The scope of process safety largely deals with fires, explosions, and toxic releases. Returning to the incident pyramid (Figure 8.1), if a management system is designed to facilitate effective process safety efforts, then investigation of near-misses must involve process-related near-misses such as temperature excursions and overpressurizations inside process vessels. Investigation of OH&S near-misses involving working at height, for example, are ineffective and may be counterproductive if the desired focus is process safety. (These measures are, of course, entirely appropriate if the focus is on OH&S.)

This distinction between occupational and process safety is far from a simple academic exercise. As noted in Chapter 1, hydrogen is widely produced and used in the chemical and oil industries (broadly speaking, the process industries) with growing use in the public realm [9]. While the former area of focus would obviously necessitate process safety considerations, the latter would involve transportation uses in which occupational safety issues for individuals are paramount.

In its document on safety plan development for DOE-funded hydrogen and fuel cell projects, the DOE addresses the above point by providing the following guidance on the first element of such a plan—the scope of work [10]:

> The plan should briefly describe the specific nature of the work being performed to set the context for the safety plan. It should distinguish between laboratory-scale research, bench-scale testing, engineering development, and prototype operation. All intended project phases should be described.

Implicit in the above guidance is the need to identify the hazards of concern and the affected personnel when establishing the scope of work to be addressed by the safety plan (which is in essence a description of a safety management system). Interestingly, the section on scope of work [10] goes on to describe the value of quantifying the amounts of hazardous materials (including hydrogen) generated, used, and stored, thus emphasizing the linkage between inherently safer design (Chapter 7) and effective safety management. The complete set of DOE safety plan elements is given in Table 8.1; further discussion of this framework is undertaken in the next section.

8.2 Process Safety Management

Having established the prevalence of hydrogen use in the process industries, as well as the clear need to prevent and mitigate process-related incidents, it is appropriate to now turn solely to the matter of process safety management. To reiterate, process safety management deals with the identification, understanding, and control of process hazards to prevent process-related injuries and incidents (fires, explosions, toxic releases). As previously mentioned, the discussion here is based on Kletz and Amyotte [3], with specific reference to hydrogen safety.

TABLE 8.1

Safety Plan Elements for U.S. DOE-Funded Hydrogen and Fuel Cell Projects

No.	Element
1	Scope of work
2	Organizational safety information
	Organizational policies and procedures
	Hydrogen and fuel cell experience
3	Project safety
	Identification of safety vulnerabilities
	Risk reduction plan
	Operating procedures
	Equipment and mechanical integrity
	Management of change procedures
	Project safety documentation
4	Communication plan
	Employee training
	Safety reviews
	Safety events and lessons learned
	Emergency response
	Self-audits
5	Safety plan approval
6	Other comments or concerns

Source: DOE, *Safety Planning Guidance for Hydrogen and Fuel Cell Projects*, Fuel Cell Technologies Program, U.S. Department of Energy (April 2010).

In his recounting of the history of safety philosophy in the process industries, Creedy [7] describes four phases of development.

1. The late nineteenth and early twentieth century: Here the objective was primarily the protection of capital assets, providing the origins of basic safety thinking in the explosives industry.
2. The Second World War through the 1950s and 1960s: During this period the objectives were greater efficiency and the creation of a better society. The concepts of loss prevention and investing in people were introduced. Safety measures were largely rule based.
3. The 1970s and 1980s: The objectives here were the same as the previous phase. However, recognition of consequence seriousness and causation mechanisms led to a focus on the process rather than the individual worker, and hence the development of a management approach to process safety (*process* being defined by the U.S. Occupational Health and Safety Administration as any activity

involving a highly hazardous chemical, including any use, storage, manufacturing, handling, or the on-site movement of such chemicals, or combination of these activities).

4. The 1990s and beyond: Here we see a realization of the significance of sociocultural factors in human thought processes and behaviors at the individual and organizational levels. This has led to increased understanding of the importance of such concepts as human factors and safety culture.

At present, we appear to be in a period that merges the last two of the above phases—process safety management coupled with a recognition that without a strong safety culture (to be defined later in this chapter), even the best management system on paper can become dysfunctional. An approach widely used in Canada is known simply as PSM (Process Safety Management). The complete suite of PSM elements is shown in Table 8.2, taken from the Process Safety Management Guide of the Canadian Society for Chemical Engineering (CSChE) [11].

This guide was prepared by the Process Safety Working Group of the former Major Industrial Accidents Council of Canada (MIACC) in conjunction with the Process Safety Management Committee of the Canadian Chemical Producers' Association (CCPA), now known as the Chemistry Industry Association of Canada. With the dissolution of MIAC in 1999, rights to the guide were transferred to the CSChE. The material in the CSChE PSM guide [11] is based on that developed by the Center for Chemical Process Safety (CCPS) of the American Institute of Chemical Engineers (e.g., [12]). This route was adopted because the CCPS approach to process safety management was determined to be comprehensive, well-supported by reference materials, tools, and an organizational structure, and based on a benchmark of leading or good industry practice rather than on minimum standards [11].

Table 8.2 will therefore likely be familiar to PSM practitioners throughout North America. It will also be relevant to those engaged in process safety efforts in other parts of the world given the incorporation of best practices in the elements and components shown in Table 8.2. Other systems may have more or fewer elements, the terminology may be somewhat different, or a specific management system may be mandated by regulation, but the underlying concepts outlined in the previous section are the same.

As previously discussed, continuous improvement is a key feature of safety management systems; in this regard it is important to note the recent CCPS work on developing a framework for risk-based process safety (RBPS) management [13]. This 20-element, risk-based system is shown in Table 8.3. As an evolution of the PSM approach given in Table 8.2, the RBPS management system makes an explicit link between process safety *management* and process safety *culture*.

TABLE 8.2

Elements and Components of Process Safety Management

No.	Element	Component
1	Accountability: objectives and goals	1.1 Continuity of operations; 1.2 Continuity of systems; 1.3 Continuity of organization; 1.4 Quality process; 1.5 Control of exceptions; 1.6 Alternative methods; 1.7 Management accessibility; 1.8 Communications; 1.9 Company expectations
2	Process knowledge and documentation	2.1 Chemical and occupational health hazards; 2.2 Process definition/design criteria; 2.3 Process and equipment design; 2.4 Protective systems; 2.5 Normal and upset conditions (operating procedures); 2.6 Process risk management decisions; 2.7 Company memory (management of information)
3	Capital project review and design procedures	3.1 Appropriation request procedures; 3.2 Hazard reviews; 3.3 Siting; 3.4 Plot plan; 3.5 Process design and review procedures; 3.6 Project management procedures and controls
4	Process risk management	4.1 Hazard identification; 4.2 Risk analysis of operations; 4.3 Reduction of risk; 4.4 Residual risk management; 4.5 Process management during emergencies; 4.6 Encouraging client and supplier companies to adopt similar risk management practices; 4.7 Selection of businesses with acceptable risk
5	Management of change	5.1 Change of process technology; 5.2 Change of facility; 5.3 Organizational changes; 5.4 Variance procedures; 5.5 Permanent changes; 5.6 Temporary changes
6	Process and equipment integrity	6.1 Reliability engineering; 6.2 Materials of construction; 6.3 Fabrication and inspection procedures; 6.4 Installation procedures; 6.5 Preventative maintenance; 6.6 Process, hardware and systems inspection and testing; 6.7 Maintenance procedures; 6.8 Alarm and instrument management; 6.9 Decommissioning and demolition procedures
7	Human factors	7.1 Operator-process/equipment interface; 7.2 Administrative control versus hardware; 7.3 Human error assessment
8	Training and performance	8.1 Definition of skills and knowledge; 8.2 Design of operating and maintenance procedures; 8.3 Initial qualifications assessment; 8.4 Selection and development of training programs; 8.5 Measuring performance and effectiveness; 8.6 Instructor program; 8.7 Records management; 8.8 Ongoing performance and refresher training
9	Incident investigation	9.1 Major incidents; 9.2 Third party participation; 9.3 Follow-up and resolution; 9.4 Communication; 9.5 Incident recording, reporting and analysis; 9.6 Near-miss reporting
10	Company standards, codes and regulations	10.1 External codes/regulations; 10.2 Internal standards
11	Audits and corrective actions	11.1 Process safety management systems audits; 11.2 Process safety audits; 11.3 Compliance reviews; 11.4 Internal/external auditors
12	Enhancement of process safety knowledge	12.1 Quality control programs and process safety; 12.2 Professional trade and association programs; 12.3 CCPS program; 12.4 Research, development, documentation and implementation; 12.5 Improved predictive system; 12.6 Process safety resource centre and reference library

Source: *Process Safety Management*, 3rd edition, Canadian Society for Chemical Engineering, Ottawa, Ontario, 2002. With permission.

TABLE 8.3

Risk-Based Process Safety Management System

Accident Prevention Pillar	Risk-Based Process Safety Element
Commit to process safety	Process safety culture
	Compliance with standards
	Process safety competency
	Workforce involvement
	Stakeholder outreach
Understand hazards and risk	Process knowledge management
	Hazard identification and risk analysis
Manage risk	Operating procedures
	Safe work practices
	Asset integrity and reliability
	Contractor management
	Training and performance assurance
	Management of change
	Operational readiness
	Conduct of operations
	Emergency management
Learn from experience	Incident investigation
	Measurement and metrics
	Auditing
	Management review and continuous improvement

Source: CCPS, *Guidelines for Risk Based Process Safety*, John Wiley & Sons, Hoboken, NJ, 2007. With permission.

The generic safety management systems shown in Tables 8.2 and 8.3 share many commonalities with the hydrogen-specific safety plan given in Table 8.1. The implication here is that classical process safety concepts and methodologies are entirely applicable to the production, storage, and use of hydrogen. This is especially the case for *process risk management* (element 4 in Table 8.2) as illustrated later in this section. A specific example involving element 6 in Table 8.2 will serve to further illustrate the validity of this claim. This element, expressed in slightly different terms meaning the same thing, is as follows:

- Table 8.1, equipment and mechanical integrity
- Table 8.2, process and equipment integrity
- Table 8.3, asset integrity and reliability

The former CCPA Process Safety Management Committee has collected and analyzed data on an annual basis for process-related incidents reported by then CCPA member companies using a procedure known as PRIM (Process-Related Incidents Measure). The 2004 PRIM analysis of 89 reported incidents

TABLE 8.4

Incident Causation (2004 PRIM Data) According to PSM Elements Given in Table 8.2

No.	Element	Percent of Incidents
6	Process and equipment integrity	23.8
2	Process knowledge and documentation	21.2
4	Process risk management	16.8
7	Human factors	8.9
5	Management of change	7.3
3	Capital project review and design procedures	6.5

Source: Amyotte, P.R. et al., *Process Safety Progress*, 26 (4), 333–346, 2007. With permission.

demonstrated that six of the PSM elements in Table 8.2 contributed to 85 percent of the total incidents [14]. As shown in Table 8.4, deficiencies in *process and equipment integrity* (element 6 in Table 8.2) contributed to approximately 24 percent of the total reported incidents. Figure 8.2 shows a breakdown of this PSM element into its components (or subelements) for 2004 as well as the five reporting periods prior to 2004. Here we see the predominant role of *preventative maintenance* (component 6.5) and *maintenance procedures* (component 6.7). This observation is consistent with the comment made earlier in this chapter with respect to inadequate permit controls for maintenance activities being one of the root causes of the 1989 Pasadena, Texas, vapor cloud explosion involving hydrogen (and other hazardous materials).

The PRIM methodology uses multiple analysts (i.e., self-reporting by companies) with overall review by a team of process safety experts. It is therefore best to draw only broad conclusions as to the relative importance of particular PSM elements, especially with respect to trend analysis from year to year. For example, it seems reasonable to conclude from Figure 8.3 that for the seven-year reporting period, deficiencies in PSM elements 2 to 7 (Table 8.2) inclusive have been viewed as key contributors to process-related incidents in Canadian process industry companies. The importance attached to each of these six elements has varied over this period, but each has crossed an arbitrary threshold of having contributed to at least 10 percent of the total incidents during a given year (at least once during the seven-year period).

Further validation of the conclusions reached in the preceding paragraph is provided by Figure 8.4. This comes from the work of Amyotte, MacDonald, and Khan [15], who reviewed approximately 60 investigation reports produced by the U.S. Chemical Safety Board (CSB); the review was conducted from a PSM perspective, looking for examples of application (or lack thereof) of safety measures categorized by the hierarchy of controls—inherent, passive engineered, active engineered, and procedural (see Chapter 7). Here again we see the importance of *process and equipment integrity* (element 6 in Table 8.2, or

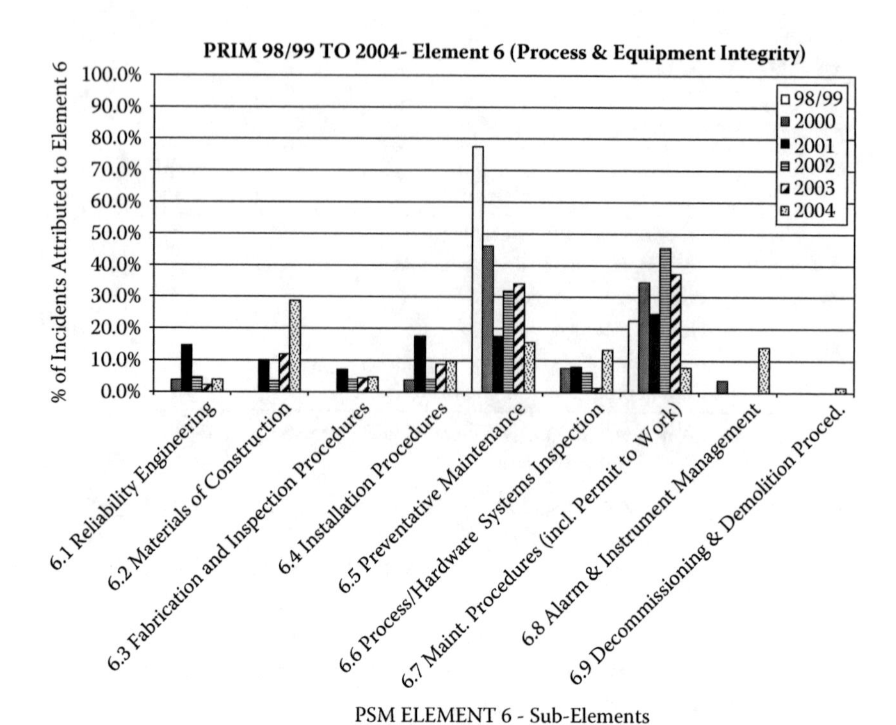

FIGURE 8.2
Incident causation (PRIM data) according to PSM element 6 given in Table 8.2. (From Amyotte, P.R. et al., *Process Safety Progress*, 26 (4), 333–346, 2007. With permission.)

asset integrity and reliability in the U.S. DOE hydrogen safety plan [10] elements shown in Table 8.1). Included in the bar for PSM element 6 in Figure 8.4 is the ISD simplification example [16] described in Chapter 7 involving a hydrogen release and subsequent fire due to incompatible materials of construction. Figure 8.4 also shows that PSM elements 2 to 6 (Table 8.2) inclusive are again dominant, as in Figure 8.3. The role of element 8 (*training and performance*) in Figure 8.4 is heightened over that in Figure 8.3 because of the large number of procedural safety measures identified in the CSB reports reviewed.

While the PRIM type of analysis does offer a method for prioritized resource allocation aimed at improvement, opportunities for risk reduction will be missed if all management system elements are not examined. For example, it is not surprising that *accountability: objectives and goals* (element 1 in Figure 8.3) is rated lower than other elements as an incident causation factor in a system of self-reporting by engineers who may be predisposed to finding technical solutions. *Incident investigation* (element 9 in Figure 8.3) might be thought of as being reactive and therefore of limited use as a preventive measure; this is an erroneous conclusion. With respect to *enhancement of process safety knowledge* (element 12 in Figure 8.3), it can be difficult to know what you do not know.

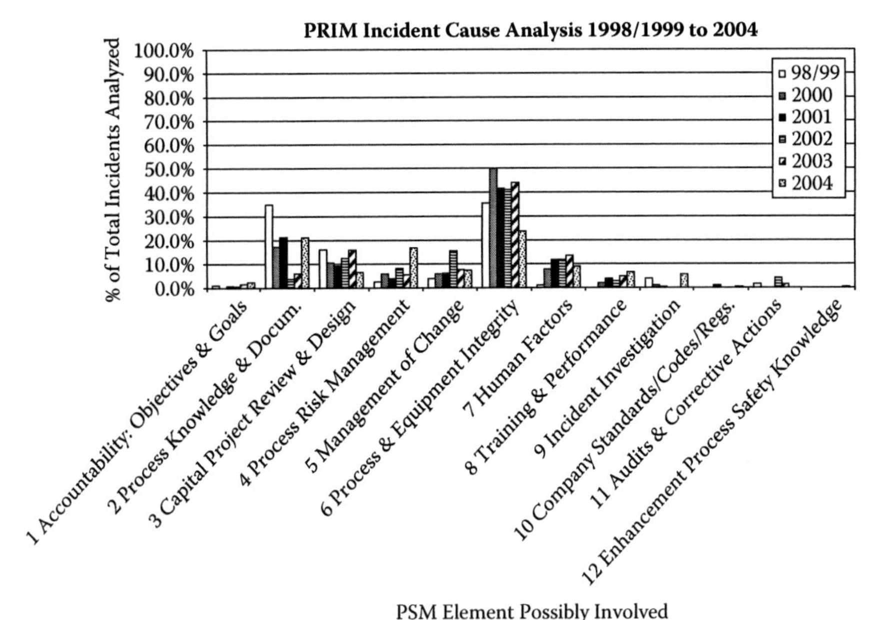

FIGURE 8.3
Incident causation (PRIM data) according to PSM elements given in Table 8.2. (From Amyotte, P.R. et al., *Process Safety Progress*, 26 (4), 333–346, 2007. With permission.).

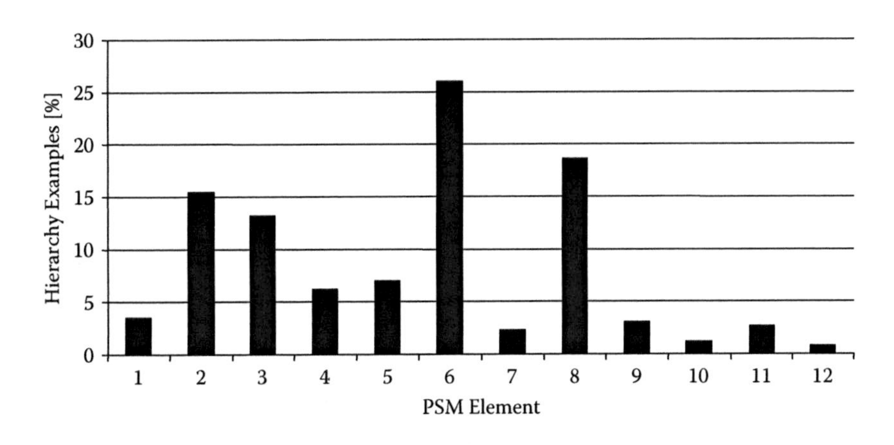

FIGURE 8.4
Breakdown of hierarchy of risk control examples according to applicable PSM elements given in Table 8.2. (From Amyotte, P.R. et al., *Process Safety Progress*, 30 (3), 261–265, 2011. With permission.)

With these points in mind, the suite of PSM elements referenced in Table 8.2 and Figure 8.3 is now examined for its application to hydrogen safety via the system elements given in Table 8.1 and other relevant examples.

8.2.1 PSM Element 1—Accountability: Objectives and Goals

> Management commitment at all levels is necessary for PSM to be effective. The objectives for establishing accountability are to demonstrate the status of process safety compared to other business objectives (e.g. production and cost), to set objectives for safe process operation and to set specific process safety goals. These objectives should be internally consistent [11].

The fundamental underpinning of an effective process safety management system is the belief that safety is a corporate value and that performance improvement requires leadership for breakthrough results [17]. This issue of whether a company believes it is possible to achieve a higher standard of safety has been addressed in a recent book by sociologist Andrew Hopkins of the Australian National University. Hopkins [18] describes three concepts that address a company's cultural approach to safety, and makes the argument that the three are essentially alternative ways of talking about the same phenomena.

1. *Safety culture.* The concept of a safety culture embodies the following subcultures [18]:
    - A *reporting culture* in which people report errors, near-misses, substandard conditions, inappropriate procedures, etc.
    - A *just culture* in which blame and punishment are reserved for behavior involving defiance, recklessness, or malice, such that incident reporting is not discouraged.
    - A *learning culture* in which a company learns from its reported incidents, processes information in a conscientious manner, and makes changes accordingly.
    - A *flexible culture* in which decision-making processes are not so rigid that they cannot be varied according to the urgency of the decision and the expertise of the people involved.
2. *Collective mindfulness.* The concept of collective mindfulness embodies the principle of *mindful organizing,* which incorporates the following processes [18]:
    - A *preoccupation with failure* so that a company is not lulled into a false sense of security by periods of success. A company that is preoccupied with failure would have a well-developed reporting culture.

- A *reluctance to simplify* data that may at face value seem unimportant or irrelevant, but which may in fact contain the information needed to reduce the likelihood of a future surprise. (Note that simplification here is not a desirable goal, unlike the ISD principle of simplification.)
- A *sensitivity to operations* in which frontline operators and managers strive to remain as aware as possible of the current state of operations, and to understand the implications of the present situation for future functioning of the company.
- A *commitment to resilience* in which companies respond to errors or crises in a manner appropriate to deal with the difficulty, and a *deference to expertise* in which decisions are made by the people in the company hierarchy who have the most appropriate knowledge and ability to deal with the difficulty.

3. *Risk awareness.* Hopkins [18] states that risk-awareness is synonymous with collective mindfulness (which is obviously closely related to the concept of a safety culture). He also describes a culture of *risk denial* in which it is not simply a matter of individuals and companies being unaware of risks, but rather that there exist mechanisms that deny the existence of risk.

From an overall perspective, element 2 (*organizational safety information*) in the DOE hydrogen safety plan (Table 8.1) pertains to this PSM element, particularly the setting of *organizational policies and procedures.* Management accountability must also permeate the entire safety management system. This is demonstrated by the relevance of Hopkin's points on safety culture [18] to the element 4 component *safety events and lessons learned* (Table 8.1). If a just culture does not exist, incident reporting will be sporadic at best and there will be no lessons that can be learned from nonexistent investigation reports. The concepts of collective mindfulness and risk awareness [18] can be seen to relate directly to several of the element 3 components in Table 8.1, for example, *identification of safety vulnerabilities* and *risk reduction plan.*

8.2.2 PSM Element 2—Process Knowledge and Documentation

> Information necessary for the safe design, operation and maintenance of any facility should be written, reliable, current and easily accessible by people who need to use it [11].

There is a close correspondence between this PSM element and the element 3 component *project safety documentation* (as well as *operating procedures*) in Table 8.1. The DOE hydrogen safety plan [10] contains the following requirements for *project safety documentation,* all of which relate directly to the PSM element 2 components in Table 8.2:

- Information pertaining to the technology of the project
- Information pertaining to the equipment or apparatus
- Safety systems (e.g., alarms, interlocks, detection or suppression systems)
- Safety review documentation, including identification of safety vulnerabilities
- Operating procedures (including response to deviation during operation)
- Material safety data sheets
- References such as handbooks and standards

Also relevant here is PSM component 2.7, *company memory (management of information)*. The intention of this component is to ensure that knowledge and information gained from plant experience, and which is likely to be important for the future safety of a facility, is well documented so it is not forgotten or overlooked as personnel and organizational changes occur [11]. Mannan, Prem, and Ng [19] comment that organizations lose valuable information after about 10 years because of such changes. Given the relative newness of some features of the hydrogen industry, as well as the sensitivity of public opinion to some aspects of hydrogen usage, it would seem opportune to heed the lessons experienced—sometimes with significant hardship—by more-established process industry sectors.

8.2.3 PSM Element 3—Capital Project Review and Design Procedures

Many industrial practitioners hold the opinion that careful attention to this element can have the greatest impact on the effectiveness of process safety management [20]. The key here is to conduct *hazard reviews* (PSM component 3.2 in Table 8.2) early in the design sequence by employing a preliminary hazard analysis. This is essentially the advice given in the DOE hydrogen safety plan [10] under the element 4 component *safety reviews*, which extends the review concept throughout the life cycle of a project. Life cycle considerations must, by definition, include the front end of the project.

PSM components 3.3, *siting*, and 3.4, *plot plan*, therefore take on notable significance. In siting a proposed expansion or new plant, the exposure hazard to and from adjacent plants or facilities is a critical consideration; similarly the location of control rooms, offices, and other buildings should be carefully considered in conducting a plot plan review [11]. This is in accordance with the safety discussion in Section 7.6 on avoiding knock-on or domino effects, and the use of unit segregation in the hierarchy of controls (Figure 7.1 in Chapter 7).

Well-known examples where greater attention to facility siting and plot plan review (temporary as well as capital) would have assisted with consequence mitigation include the administration and control buildings at

Flixborough [21] and the contractor trailers at the BP Texas City refinery [22]. Several hydrogen-specific examples were given in Section 7.6; for example, the work of Matthijsen and Kooi [23] in determining safety distances for hydrogen filling stations. The motivation for this work was to minimize external or third-party risk (i.e., the risk exposure of people living or working in the vicinity of facilities handling large amounts of hazardous substances) [23].

8.2.4 PSM Element 4—Process Risk Management

The PSM guide comments that component 4.1, hazard identification, is the most important step in process risk management: If hazards are not identified, they cannot be considered in implementing a risk reduction program, nor addressed by emergency response plans [11].

This is similar to the distinction between hazard and risk made by Crowl and Jo [24]; as illustrated in Figure 8.5, effective assessment of the risk components of incident probability (or likelihood) and severity of consequences can only be carried out after thorough identification of the relevant hazards. The important aspects of system description (i.e., establishing the physical and analytical scopes) and scenario identification (i.e., identifying credible scenarios for study) have been addressed by Takeno et al. [25] and Gerboni and Salvador [26]. Pasman and Rogers [27] further comment on the need to focus on prevention and mitigation measures aimed at *both* risk components so as to facilitate the smooth introduction of large-scale use of hydrogen as a transportation fuel.

Several techniques are referenced in the PSM guide [11] for identifying and assessing hazards, including what-if (WI) analysis, checklist (CL) analysis, hazard and operability (HAZOP) study, failure modes and effects analysis (FMEA), fault tree analysis (FTA), and the Dow Fire and Explosion Index (F&EI) and Chemical Exposure Index (CEI). These and a number of other hazard identification/risk assessment methodologies have been well described by the Center for Chemical Process Safety (CCPS) of the American Institute of Chemical Engineers [28]. The DOE hydrogen safety plan [10] provides similar guidance under the element 3 component *identification of safety vulnerabilities* (ISV), which lists the following ISV methods also given in the PSM guide [11]: FMEA, WI, HAZOP, CL, and FTA. Additional methods referenced include event tree analysis (ETA) and probabilistic (or quantitative) risk assessment (PRA or QRA) [10]. These techniques find application in both nonemergency situations and those requiring an *emergency response* (element 4 component in Table 8.1).

A review of the hydrogen safety literature reveals that essentially all of the typical process safety hazard identification techniques have been successfully applied to various sectors of the hydrogen industry. Knowlton [29] used HAZOP to study the safety aspects of hydrogen as a ground transportation fuel almost 30 years ago. More recently, HAZOP and FMEA were used

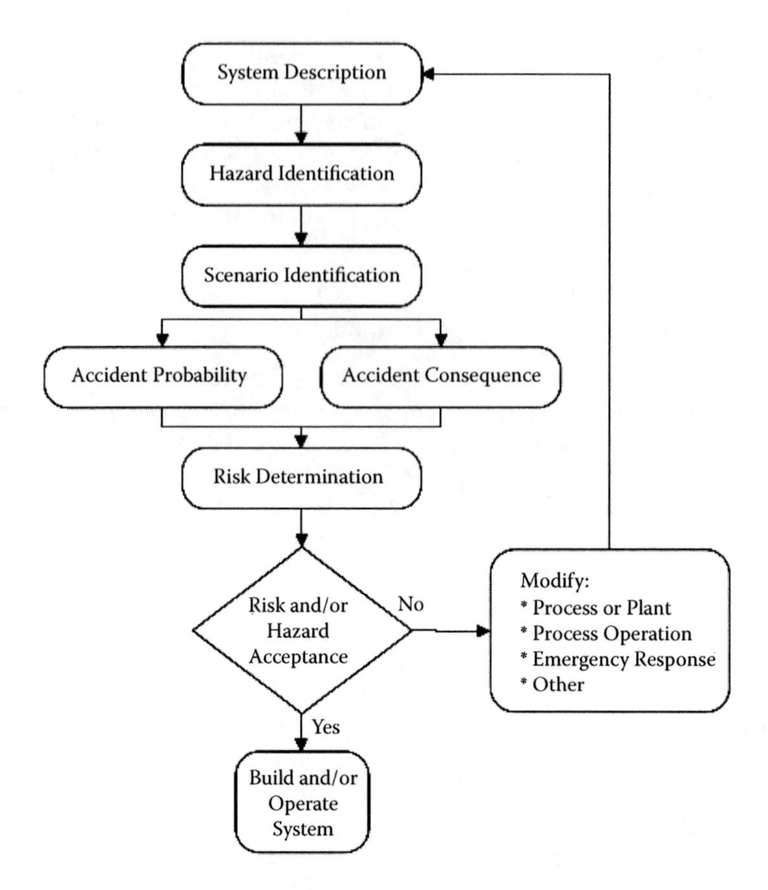

FIGURE 8.5
Risk management process involving hazard identification and risk assessment. (From Crowl, D.A., and Jo, Y.-D., *Journal of Loss Prevention in the Process Industries*, 20 (2), 158–164, 2007. With permission.)

by Kikukawa, Mitsuhashi, and Miyake [30] to identify hazards and assess the ensuing risks for liquid hydrogen fueling stations (Figure 8.6). Their risk assessment process is shown in Figure 8.7; this flow chart is similar to the general one given in Figure 8.5, with the understanding that any risk reduction measures recommended through use of Figure 8.7 should be thoroughly examined for the introduction of new hazards.

The risk matrix used by Kikukawa, Mitsuhashi, and Miyake [30] to discern whether a given risk is tolerable is shown in Figure 8.8. Risk matrices are a commonly used decision tool in the process industries and are referenced in the DOE hydrogen safety plan [10] under the element 3 component *risk reduction plan* by means of the term "risk binning matrix." Additional information on risk matrices and an associated concept, the ALARP (as low as reasonably practicable) principle, has been given for hydrogen refueling stations by Norsk Hydro ASA and DNV [31].

FIGURE 8.6
Liquid hydrogen fueling station. (From Kikukawa, S. et al., *International Journal of Hydrogen Energy*, 34 (2), 1135–1141, 2009. With permission.)

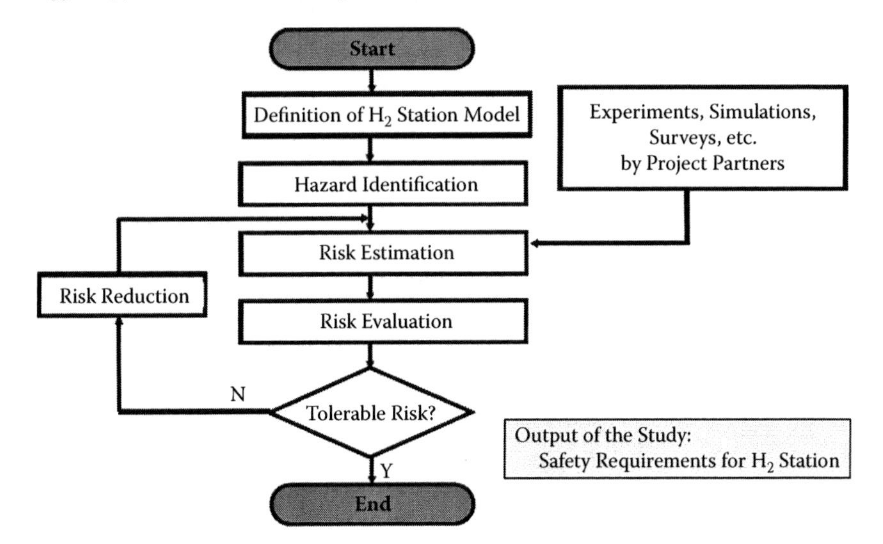

FIGURE 8.7
Risk assessment process for liquid hydrogen fueling station. (From Kikukawa, S. et al., *International Journal of Hydrogen Energy*, 34 (2), 1135–1141, 2009. With permission.)

FTA, along with HAZOP and PHA (preliminary hazard analysis), were used by Brown and Buchier [32] in their study of a hydrogen gaseous effluent treatment and purification plant. Figure 8.9 shows the fault tree developed for the top event of a hydrogen compressor explosion. In its original form (Figure 8.9), the fault tree is in essence a logic diagram with an implicit "and" gate beneath the Compressor Explosion block and implicit "or" gates beneath

Consequence Severity Level		A Improbable	B Remote	C Occasional	D Probable
		Probability Level			
1	Extremely Severe Damage	H	H	H	H
2	Severe Damage	M	H	H	H
3	Damage	M	M	H	H
4	Limited Damage	L	L	M	H
5	Minor Damage	L	L	L	M

FIGURE 8.8
Risk matrix for liquid hydrogen fueling station. H = High risk; M = medium risk; L = low risk. (From Kikukawa, S. et al., *International Journal of Hydrogen Energy*, 34 (2), 1135–1141, 2009. With permission.)

the blocks for Explosive Atmosphere Formation, H_2 Tubing Collapse, and Ignition Source. Nevertheless, Figure 8.9 does demonstrate the usefulness of such graphical techniques for identifying the potential hazards leading to an undesired event.

FTA and ETA were used by Roysid, Jablonski, and Hauptmanns [33] and Rodionov, Wilkening, and Moretto [34] in their safety studies of lifecycle-related hydrogen usage and private vehicles with a hydrogen-driven engine, respectively. The use of event trees figured prominently in the work of Gerboni and Salvador [26] on hydrogen transportation systems and that of Rigas and Sklavounos [35] on hydrogen storage facilities.

Figure 5.4 shows an event tree for the case of a hydrogen release, drawn from the study of Rigas and Sklavounos [35]. The usefulness of graphical hazard/risk techniques is again demonstrated in Figure 5.4 by the clear elucidation of the ultimate result of various mitigating factors following an undesired event. For example, a hydrogen release followed by immediate ignition in a region of significant confinement is seen to lead to either a deflagration or a detonation. On the other hand, the same release without subsequent ignition and with no confinement results in dissolution of the hydrogen plume. Risk reduction measures can then be decided upon with the aid of Figure 5.4.

The Dow Fire and Explosion Index (F&EI), often considered as a form of relative risk ranking (RRR), was used by Bernatik and Libisova [36] in conjunction with HAZOP, FTA, and ETA to conduct a risk assessment study of large gasholders in municipal areas. The F&EI is ideally suited for use with hydrogen given its high value of the U.S. National Fire Protection Association (NFPA) flammability number ($N_F = 4$); the Dow Chemical Exposure Index (CEI), however, would be of limited use with hydrogen because of the low NFPA health hazard number ($N_H = 0$). The use of RRR was also adopted by Kim, Lee, and Moon [37] by means of FMEA and FTA coupled with a relative risk index for comparison of hydrogen production, storage, and transportation activities.

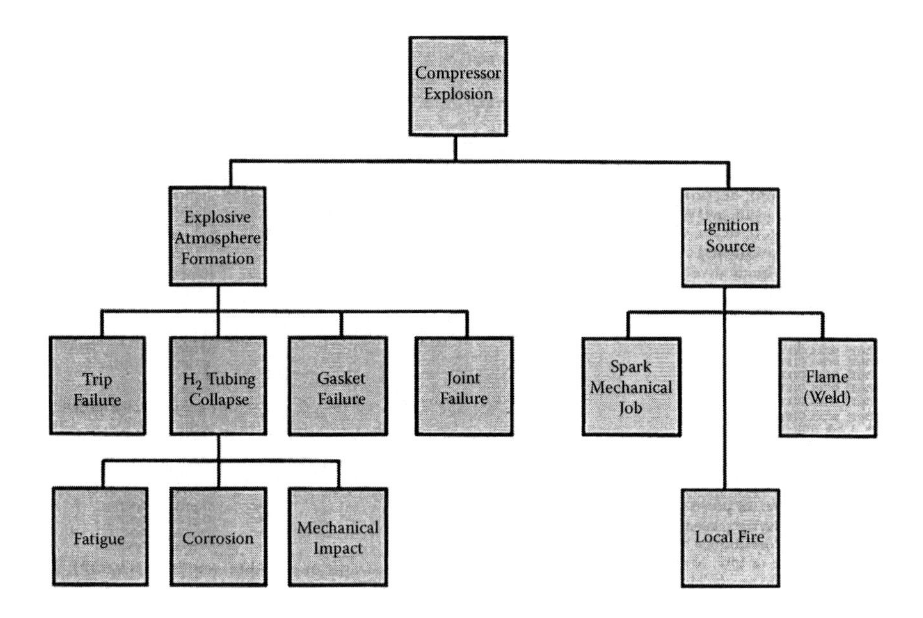

FIGURE 8.9
Fault tree analysis (FTA) for a case of hydrogen compressor explosion. (From Brown, A.E.P., and Buchier, P.M., *Process Safety Progress*, 18 (3), 166–169, 1999. With permission.)

There are also numerous reports in the hydrogen safety literature describing knowledge acquisition to better understand specific industrial hazards; again, this is a key step in the risk assessment process which first relies upon effective hazard identification. For example, Petukhov, Naboko, and Fortov [38] investigated the deflagration/detonation relationship for hydrogen-air mixtures in large volumes by conducting explosion tests in a 12-m diameter chamber. Sommersel et al. [39] undertook experiments in a 3-m long channel to gain a better understanding of hydrogen dispersion, ignition, and explosion development in such configurations.

Imamura, Mogi, and Wada [40] obtained empirical data on hydrogen ignition by electrostatic discharge at the outlet of a ventilation duct. The practical concern in this case was the possibility of ignition during routine and emergency release of hydrogen to the atmosphere through ventilation ducts such as the one shown in Figure 8.10. These researchers determined that when the ventilation duct itself was not grounded, electrostatic discharge between grounded conductors and the duct outlet, with subsequent hydrogen ignition, was a distinct possibility. Grounding of the ventilation duct outlet effectively eliminated this problem, thus demonstrating the efficacy of grounding as a risk reduction measure in this particular application [40].

Quantification of one or both of the components of risk (incident likelihood and consequence severity) has become increasingly popular in the process industries, especially as more reliable failure frequency databases and computational tools for consequence analysis have become available. The

FIGURE 8.10
Ventilation duct outlet at hydrogen fueling station in Japan. (From Imamura, T. et al., *International Journal of Hydrogen Energy*, 34 (6), 2815–2823, 2009. With permission.

same statement can be made with confidence for the hydrogen industry as evidenced by the proliferation of articles describing the use of: (1) the Dutch Purple Book for determining risk scenarios and failure frequencies [23, 36], (2) semiquantitative risk assessment methodologies [41], (3) probit functions for translating radiant heat flux calculations to probabilities of first- and second-degree burns as well as fatalities [42], (4) Gaussian dispersion modeling [43], (5) Bayesian networks as a supplement to FTA and ETA [44], and (6) QRA as the foundation for making risk-based management decisions [45].

It can be argued that in this regard, some of the most important developments in recent years have been in the field of computational fluid dynamics (CFD). The GexCon CFD-tool FLACS (FLame ACceleration Simulator), for example, has been adapted for use in the hydrogen industry to simulate accidental releases and dispersion of hydrogen followed by ignition and flame and overpressure development [46, 47]. Further investigation with FLACS has been reported for a wide range of hydrogen applications in various natural and manufactured environments [48–55]; these studies have been carried out within the framework of the European Commission-funded Network of Excellence *HySafe* (Hydrogen Safety as an Energy Carrier), which is briefly described in Chapter 9.

Concluding comments can be made for this PSM element with respect to component 4.6, *encouraging client and supplier companies to adopt similar risk management practices* (Table 8.2). This component is particularly important because of the common practice of outsourcing or contracting of engineering services. This is often the case on large projects where partnerships and joint ventures are formed, but may also apply to smaller projects through the awarding of subcontracts. Success on these projects is in large measure determined by the degree of commonality in risk management practices among the different parties. Not the least of the concerns is whether there is a common set of expectations for safety performance and risk-awareness. The DOE hydrogen safety plan [10] addresses these points under element 1, *scope of work* (Table 8.1), with reference to the need for such safety plans to cover the work of all subcontractors.

8.2.5 PSM Element 5—Management of Change

> A system to manage change is critical to the operation of any facility. A written procedure should be required for all changes except replacement-in-kind. The system should address: a clear definition of change (scope of application); a description and technical basis for the proposed change; potential impact of the proposed change in health, safety and environment; authorization requirements to make the change; training requirements for employees or contractors following the change; updating of documentation including process safety information, operating procedures, maintenance procedures, alarm and interlock settings, fire protection systems, etc.; and contingencies for "emergency" changes [11].

There is clearly a direct correspondence between this PSM element and the DOE hydrogen safety plan [10] element 3 component, *management of change procedures* (Table 8.1). The authors of this plan [10] have seemingly heeded the advice of Hansen and Gammel [56], who state that while management of change (MOC) is critical to process safety, it is a concept that, if well implemented, could help prevent accidents in many other industries as well.

MOC is of particular importance when employing the inherently safer design principles described in Chapter 7. Simply put, inherent safety involves change, and change in any industrial sector must be managed. Potential hazards brought about by inherent safety changes must therefore be identified and the ensuing risk reduced to an acceptable level. This is as important as the concept of looking for inherent safety opportunities when making a process change. For example, the development of nano-porous carbon tubes and their substitution for liquid hydrogen storage [9] would necessitate careful consideration of new hazards that might be introduced through the use of nanomaterials.

8.2.6 PSM Element 6—Process and Equipment Integrity

> Procedures for fabricating, inspecting and maintaining equipment are vital to process safety. Written procedures should be used to maintain ongoing integrity of process equipment such as: pressure vessels and storage tanks; piping, instrumentation and electrical systems; process control software; relief and vent systems and devices;

> emergency and fire protection systems; controls including monitoring devices and sensors, alarms and interlocks; and rotating equipment. A documented file should be maintained for each piece of equipment [11].

The relationship between this PSM element and the DOE hydrogen safety plan has been discussed extensively in the introductory material to this section.

An additional comment concerning *maintenance procedures* (PSM component 6.7 in Table 8.2), follows from a safety alert issued by the U.S. Environmental Protection Agency [57], which offered the advice that facilities with storage tanks containing flammable vapors (such as hydrogen) should review their equipment and operations in the following areas:

- Design of atmospheric storage tanks
- Inspection and maintenance of storage tanks
- Hot-work safety
- Ignition source reduction

While the middle two items in this list are largely procedural in nature, the first and last items have inherent safety overtones (simplification and minimization, respectively). This again demonstrates the linkage between the subject matter of the current chapter (safety management systems) and Chapter 7 (inherently safer design).

8.2.7 PSM Element 7—Human Factors

> Human factors are a significant contributor to many process accidents. Three key areas are operator–process/equipment interface, administrative controls and human error assessment [11].

Human error and the underlying human factors have been recognized as causation factors for several industrial accidents involving hydrogen [58]. Xu et al. [59] present data indicating that for a subset of such accidents occurring over the period 1972 to 2005, human error was a root cause in 26 percent of the total number of incidents (with management system deficiencies accounting for a further 27 percent). It is therefore surprising that the DOE hydrogen safety plan [10] does not contain a separate element or component to explicitly address human factors. This is clearly one area where hydrogen safety management can benefit from the lessons of process safety management in general.

The PSM element on human factors has a strong relationship with the principles of inherently safer design (ISD), particularly simplification (Chapter 7).

Component 7.1, *operator-process/equipment interface* (Table 8.2), refers to issues such as [11]:

- The design of equipment increasing the potential for error (e.g., confusing equipment, positioning of dials, color coding, different directions for on/off, etc.).
- The need for a task analysis (a step-by-step approach to examine how a job will be done) to determine what can go wrong during the task and how potential problem areas can be controlled.

These issues are applicable to equipment and procedures involved with hydrogen usage as well as other hazardous materials.

The description of component 7.2 of this PSM element, *administrative control versus hardware control* (Table 8.2) includes:

> Hazards may be controlled by the use of procedures or by the addition of protective equipment. This balance is often a matter of company culture and economics. If procedures are well understood, kept current and are used, then they are likely to be effective. Similarly protective systems need regular testing and maintenance to be effective. The problem of administrative versus hardware controls should be considered and a balance selected by conscious choice rather than allowing it to happen by default [11].

This description may leave some readers with the unfortunate impression that only procedural (administrative) and engineered (add-on) measures are available, or are effective, for hazard control. As demonstrated in Chapter 7, these categories in the hierarchy of controls are indeed helpful in facilitating hydrogen safety; the predominant effectiveness of ISD should not, however, be ignored when attempting to combat human error.

The third and final component of human factors, 7.3 *human error assessment* (Table 8.2), is perhaps one of the more challenging areas of process safety management. Human error assessment is becoming increasingly important in industry and is a growing area of concern for the public and for regulators. This is especially the case for the hydrogen industry given the findings previously described for human error as a key incident causation factor [58, 59].

8.2.8 PSM Element 8—Training and Performance

> People need to be trained in the right skills and to have ongoing retraining to maintain these skills [11].

As with MOC (Section 8.2.5), there is a clear and direct correspondence between this PSM element and the DOE hydrogen safety plan [10]—here, with the element 4 component *employee training* (Table 8.1). This helps to explain, in part, the rationale behind initiatives such as the development of a curriculum for an online hydrogen fuel training program [60].

Embodied within the components of this PSM element (Table 8.2) is the management cycle comprised of *plan, do, check, act* (Section 8.1). The activities in the training protocol given by DiBerardinis [61] are particularly appropriate in this regard:

- Conducting a needs analysis.
- Setting learning objectives.
- Deciding on the method of presentation and delivery of training.
    - Styles: lectures, group discussion, hands-on exercises, self-learning, etc.
    - Materials: overheads, slides, videos, workbooks, etc.
- Evaluating the instruction.
- Providing feedback on the instruction.

Excellent advice is given by Felder and Brent [62] concerning the setting of instructional (learning) objectives. They comment that in setting such objectives, there are four leading verbs that should be avoided: *know, learn, appreciate*, and *understand*. Thus, while it would be desirable for plant employees to know, learn, appreciate, and understand various issues related to hydrogen safety, these are not valid instructional objectives because it is not possible to directly see whether they have been done. It is necessary to consider what trainees should be asked to *do* to demonstrate their knowledge, learning, appreciation, and understanding of hydrogen safety, and then make those activities the instructional objectives [62].

Felder and Brent [62] further describe the concept of using action verbs to set instructional objectives. Using their breakdown according to *Bloom's Taxonomy of Educational Objectives*, examples of instructional objectives for hydrogen safety training would be:

- Knowledge: *List* the key material hazards of hydrogen.
- Comprehension: *Explain* in your own words the concept of moderation as applied to inherently safer design issues for hydrogen.
- Application: *Calculate* the Dow Fire and Explosion Index for several hydrogen inventory scenarios.
- Analysis: *Identify* the safety features in a given design for hydrogen storage using the hierarchy of risk control measures as a guide.

- Synthesis: *Develop* an original case study involving some aspect of hydrogen safety.
- Evaluation: *Select* from available techniques for assessing the risk of a given design for a hydrogen infrastructure project and justify your choice.

(NOTE: Although Bloom's Taxonomy has been revised somewhat in recent years, the original form as indicated above has been retained here for consistency with the work of Felder and Brent [62].)

8.2.9 PSM Element 9—Incident Investigation

The element 4 component *safety events and lessons learned* (Table 8.1) is the DOE hydrogen safety plan [10] counterpart to this PSM element. As discussed in Section 8.2.1, three of the safety subcultures (just, reporting, and learning) identified by Hopkins [18] relate directly to the issue of incident investigation. This illustrates another key point about safety management systems: the various elements and components comprising these systems are not stand-alone modules having little interaction among them.

The process safety literature contains numerous descriptions of how to effectively investigate industrial accidents and near-misses. For example, Goraya, Amyotte, and Khan [63] developed an inherent safety-based incident investigation methodology that is easily adaptable as a basic protocol for investigating hydrogen-related incidents. Key features of their work include the use of [63]:

- An overall framework having best-practice industry consensus.
- An integrated approach in considering all potential categories of loss (people, assets, business operation, and the environment).
- Evidence classification into data categories (position, people, parts, and paper).
- An established loss causation model for identification of causal factors.
- A layered investigation approach for making recommendations of an immediate technical nature, to remove the underlying hazards, and to improve the safety management system.

8.2.10 PSM Element 10—Company Standards, Codes, and Regulations

> A management system is needed to ensure that the various internal and external published guidelines, standards and regulations are current, disseminated to appropriate people and departments, and applied throughout the plant [11].

As noted above, the PSM guide [11] breaks this element into two components: external to the company and internal to its operations. External codes and regulations include legislated items; internal standards are comprised of items such as design principles and standard operating procedures. The DOE hydrogen safety plan [10] addresses these points, in part, under the element 3 component *project safety documentation* (Table 8.1) with reference to documenting the design codes and standards employed for equipment or apparatus.

From a process industry perspective, codes and standards for assuring hydrogen safety would be addressed in the first instance by the relevant process safety management regulatory regime (e.g., the highly regulated and prescriptive requirements promulgated in the United States). In other sectors of the broader hydrogen industry (e.g., transportation), considerable efforts are underway worldwide to develop risk-informed codes and standards with appropriate stakeholder input [27, 45].

8.2.11 PSM Element 11—Audits and Corrective Actions

> The purpose of safety audits is to determine the status and effectiveness of safety management efforts versus goals and also the progress toward those goals [11].

The element 4 component *self-audits* (Table 8.1) is the DOE hydrogen safety plan [10] counterpart to this PSM element (which is arguably one of the more self-explanatory elements given in Table 8.2). Although the term "self-audit" is used in the DOE plan [10], the supporting documentation does indicate the need for verification of the audit findings by a third party external to the project.

8.2.12 PSM Element 12—Enhancement of Process Safety Knowledge

> A management system for process safety should be designed for continuous improvement. Safety requirements are becoming more stringent, while knowledge of systems and technology is growing, e.g. consequence modelling techniques. Safe operation of a process plant calls for personnel to stay abreast of current developments, and for safety information to be readily accessible [11].

This PSM element is partially addressed in the DOE hydrogen safety plan [10] by the element 2 component *hydrogen and fuel cell experience* (Table 8.1). Enhancement of knowledge, perhaps more than any other element, is where the continuous improvement aspect of a safety management system should be highly evident. The following example illustrates this point.

Frank [64] asks the following question of urban fire department personnel who, although well-aware of the hazards in tenement building fires, may go many years without responding to a fire in a power plant: *How do you find out that power plants use hydrogen?* His answers are to: (1) tour the plant regularly, (2) ask where hydrogen is stored in bulk, (3) ask to be shown the location of the hydrogen lines and the hydrogen-cooled generator, and (4) think about the required actions should an incident occur [64]. While this excellent advice relates to many PSM elements, for example, those dealing with hazard identification, emergency response, training, etc., it also aligns well with the intent of the quote from the PSM guide [11] given at the start of this section.

8.3 Safety Culture

Having established the link between PSM and hydrogen safety, some thoughts are now given on the role of safety culture in today's industrial world. As identified earlier for previous sections of the current chapter, the discussion on safety culture follows that given by Kletz and Amyotte [3].

There can be no doubt that safety culture is an important topic, especially since the BP Texas City incident in 2005 [65]. Recent emphasis has also been placed on safety culture in other fields (e.g., occupational health and safety [66]) and in other applications (e.g., offshore safety [67]). Section 8.2.1 has addressed safety culture from the perspective of management leadership and accountability within a process safety management system. Table 8.3 concerning risk-based process safety (RBPS) management [13] lists the first of its four RBPS pillars as committing to process safety, and emphasizes the need to develop and sustain a culture that embraces process safety.

It is beyond the scope of this book to provide an extensive review of the literature on process safety culture. Here it is sufficient to say that two key points have emerged:

- Typical occupational safety indicators such as lost-time injuries (LTIs) are inappropriate as a primary indicator with respect to process safety (as previously discussed in Section 8.1).
- Leading indicators are generally viewed as being more useful than lagging indicators.

Similar to the development of methods to measure the inherent "safeness" of a process (Section 7.7), safety culture metrics is currently an area of significant interest in both academia and industry. According to Hopkins [68], perhaps the most important consideration for process safety indicators is that they measure the effectiveness of the various controls comprising the risk control system. This affords an opportunity to link back to the elements making up the process safety management system. Little work appears to have been done in this area, which has been identified by Glendon [69] as a major challenge for industry (i.e., the linking of safety culture methodologies with process safety approaches and broader systems safety and risk management concepts).

The coming years will undoubtedly see advances in the area of metrics for process safety culture; this will be a welcome development, especially where the measurement tools specifically address hydrogen safety. Perhaps, however, there is something to be gained by looking to the past as well as to the future. It would seem that while it has not necessarily been named as such, safety culture has been a subject of consideration for centuries.

One of the first recorded accounts of a dust explosion was written by Count Morozzo, who gave a detailed account of an explosion in a flour warehouse in Turin, Italy [70]. In the final paragraph of his report, the Count writes:

> Ignorance of the fore-mentioned circumstances, and a culpable negligence of those precautions which ought to be taken, have often caused more misfortunes and loss than the most contriving malice. It is therefore of great importance that these facts should be universally known, that public utility may reap from them every possible advantage [71].

The above passage makes an eloquent case for the importance of incident investigation and the sharing of lessons learned (Section 8.2.9), and for a strong safety culture. It is instructive to also note that it was written over 200 years ago.

Additional insight into the importance of safety culture for the hydrogen industry can be gained by looking at other areas of application. In addressing the role of safety culture in the nanotechnology field, Amyotte [72] made the following comments:

> The nanotechnology world does not want, and neither should it need, a Bhopal, Buncefield or Gulf oil leak (all major and/or recent process/environmental incidents) to drive its safety culture. Simply put, nanotechnology industries cannot afford to ignore the hard safety lessons that have been learned and at times ignored by the chemical process industries [72].

These comments apply equally well to industries involved in the production, distribution, storage, and use of hydrogen. As described in Chapter 10 and elsewhere in this book, there have been numerous industrial accidents involving hydrogen. The avoidance of further incidents will be accomplished by successful implementation of many factors—chief among which must be acknowledgment of the importance of, and adoption of specific measures to ensure, a sound safety culture.

The hydrogen industry would be well advised to look to the social and management sciences for advice on how to avoid complacency and how best to heed the warning signs that always precede industrial accidents. These signs may appear to be vague and ambiguous, and may invite what is known as *normalization of deviance* (wherein abnormal situations become accepted as the norm) [73]. The study of high-reliability organizations [73] appears to hold much promise in providing widely applicable lessons on safety culture—lessons that would be beneficial to the various sectors of the hydrogen industry.

References

1. New integrated management system attempts to link environment, health, safety and process management, *Workplace Environment Health & Safety Reporter*, 7 (1), 1166, 2001.
2. Amyotte, P.R., and McCutcheon, D.J., *Risk Management: An Area of Knowledge for all Engineers*, Discussion paper prepared for Canadian Council of Professional Engineers, Ottawa, Ontario, 2006.
3. Kletz, T., and Amyotte, P., *Process Plants: A Handbook for Inherently Safer Design*, 2nd edition, CRC Press/Taylor & Francis Group, Boca Raton, FL, 2010.
4. Amyotte, P.R., and Oehmen, A.M., Application of a loss causation model to the Westray mine explosion, *Process Safety and Environmental Protection*, 80 (1), 55–59, 2002.
5. DOE, *Hydrogen Safety*, Technical Plan – Safety; Multi-Year Research, Development and Demonstration Plan, U.S. Department of Energy, pp. 3.8-1–3.8-12, 2007.
6. Stelmakowich, A., Continuous improvement, *OHS Canada*, 19 (7), 38–39, 2003.
7. Creedy, G., *Process Safety Management*, PowerPoint presentation prepared for Process Safety Management Division, Chemical Institute of Canada, Ottawa, Ontario, 2004.
8. Bird, F.E., and Germain, G.L., *Practical Loss Control Leadership*, DNV, Loganville, GA, 1996.
9. Guy, K.W.A., The hydrogen economy, *Process Safety and Environmental Protection*, 78 (4), 324–327, 2000.
10. DOE, *Safety Planning Guidance for Hydrogen and Fuel Cell Projects*, Fuel Cell Technologies Program, U.S. Department of Energy, Washington, D.C., 2010.
11. Canadian Society for Chemical Engineering, *Process Safety Management*, 3rd edition, Canadian Society for Chemical Engineering, Ottawa, Ontario, 2002.
12. CCPS, *Guidelines for Technical Management of Chemical Process Safety*, Center for Chemical Process Safety, American Institute of Chemical Engineers, New York, 1989.
13. CCPS, *Guidelines for Risk Based Process Safety*, John Wiley & Sons, Hoboken, NJ, 2007.
14. Amyotte, P.R., Goraya, A.U., Hendershot, D.C., and Khan, F.I., Incorporation of inherent safety principles in process safety management, *Process Safety Progress*, 26 (4), 333–346, 2007.

15. Amyotte, P.R., MacDonald, D.K., and Khan, F.I., An analysis of CSB investigation reports concerning the hierarchy of controls, *Process Safety Progress*, 30, 261–265, 2011.
16. CSB, *Positive Material Verification: Prevent Errors During Alloy Steel Systems Maintenance*, Safety Bulletin, No. 2005-04-B, U.S. Chemical Safety and Hazard Investigation Board, Washington, D.C., 2006.
17. Griffiths, S., *Leadership, Commitment & Accountability—The Driver of Safety Performance*, Process Safety and Loss Management Symposium, 55th Canadian Chemical Engineering Conference, Canadian Society for Chemical Engineering, Toronto, Ontario, 2005.
18. Hopkins, A., *Safety, Culture and Risk: The Organizational Causes of Disasters*, CCH Australia Limited, Sydney, Australia, 2005.
19. Mannan, M.S., Prem, K.P., and Ng, D., Challenges and needs for process safety in the new millennium, in *Proceedings of 13th International Symposium on Loss Prevention and Safety Promotion in the Process Industries*, Vol. 1, Bruges, Belgium (June 6–9), 2010, pp. 5–13.
20. Creedy, G., private communication, 2005.
21. Sanders, R.E., Designs that lacked inherent safety: Case studies, *Journal of Hazardous Materials*, 104 (1–3), 149–161, 2003.
22. CSB, *Refinery Explosion and Fire*, Investigation Report, No. 2005-04-I-TX, U.S. Chemical Safety and Hazard Investigation Board, Washington, D.C., 2007.
23. Matthijsen, A.J.C.M., and Kooi, E.S., Safety distances for hydrogen filling stations, *Fuel Cells Bulletin*, No. 11, 12–16, 2006.
24. Crowl, D.A., and Jo, Y.-D., The hazards and risks of hydrogen, *Journal of Loss Prevention in the Process Industries*, 20 (2), 158–164, 2007.
25. Takeno, K., Okabayashi, K., Kouchi, A., Nonake, T., Hashiguchi, K., and Chitose, K., Dispersion and explosion field tests for 40 MPa pressurized hydrogen, *International Journal of Hydrogen Energy*, 32 (13), 2144–2153, 2007.
26. Gerboni, R., and Salvador, E., Hydrogen transportation systems: Elements of risk analysis, *Energy*, 34 (12), 2223–2229, 2009.
27. Pasman, H.J., and Rogers, W.J., Safety challenges in view of the upcoming hydrogen economy: An overview, *Journal of Loss Prevention in the Process Industries*, 23 (6), 697–704, 2010.
28. CCPS, *Guidelines for Hazard Evaluation Procedures*, 2nd edition, Center for Chemical Process Safety, American Institute of Chemical Engineers, New York, 1992.
29. Knowlton, R.E., An investigation of the safety aspects in the use of hydrogen as a ground transportation fuel, *International Journal of Hydrogen Energy*, 9 (1–2), 129–136, 1984.
30. Kikukawa, S., Mitsuhashi, H., and Miyake, A., Risk assessment for liquid hydrogen fueling stations, *International Journal of Hydrogen Energy*, 34 (2), 1135–1141, 2009.
31. Norsk Hydro ASA, and DNV, *Risk Acceptance Criteria for Hydrogen Refuelling Stations*, European Integrated Hydrogen Project (EIHP2), Contract: ENK6-CT2000-00442 (February 2003).
32. Brown, A.E.P., and Buchier, P.M., Hazard identification analysis of a hydrogen plant, *Process Safety Progress*, 18 (3), 166–169, 1999.
33. Rosyid, O.A., Jablonski, D., and Hauptmanns, U., Risk analysis for the infrastructure of a hydrogen economy, *International Journal of Hydrogen Energy*, 32 (15), 3194–3200, 2007.

34. Rodionov, A., Wilkening, H., and Moretto, P., Risk assessment of hydrogen explosion for private car with hydrogen-driven engine, *International Journal of Hydrogen Energy*, 36 (3), 2398–2406, 2011.
35. Rigas, F., and Sklavounos, S., Evaluation of hazards associated with hydrogen storage facilities, *International Journal of Hydrogen Energy*, 30 (13–14), 1501–1510, 2005.
36. Bernatik, A., and Libisova, M., Loss prevention in heavy industry: Risk assessment of large gasholders, *Journal of Loss Prevention in the Process Industries*, 17 (4), 271–278, 2004.
37. Kim, J., Lee, Y., and Moon, I., An index-based risk assessment model for hydrogen infrastructure, *International Journal of Hydrogen Energy*, 36 (11), 6387–6398, 2011.
38. Petukhov, V.A., Naboko, I.M., and Fortov, V.E., Explosion hazard of hydrogen-air mixtures in the large volumes, *International Journal of Hydrogen Energy*, 34 (14), 5924–5931, 2009.
39. Sommersel, O.K., Bjerketvedt, D., Vaagsaether, K., and Fannelop, T.K., Experiments with release and ignition of hydrogen gas in a 3 m long channel, *International Journal of Hydrogen Energy*, 34 (14), 5869–5874, 2009.
40. Imamura, T., Mogi, T., and Wada, Y., Control of the ignition possibility of hydrogen by electrostatic discharge at a ventilation duct outlet, *International Journal of Hydrogen Energy*, 34 (6), 2815–2823, 2009.
41. Moonis, M., Wilday, A.J., and Wardman, M.J., Semi-quantitative risk assessment of commercial scale supply chain of hydrogen fuel and implications for industry and society, *Process Safety and Environmental Protection*, 88 (2), 97–108, 2010.
42. LaChance, J., Tchouvelev, A., and Engebo, A., Development of uniform harm criteria for use in quantitative risk analysis of the hydrogen infrastructure, *International Journal of Hydrogen Energy*, 36 (3), 2381–2388, 2011.
43. Ramamurthi, K., Bhadraiah, K., and Murthy, S.S., Formation of flammable hydrogen-air clouds from hydrogen leakage, *International Journal of Hydrogen Energy*, 34 (19), 8428–8437, 2009.
44. Haugom, G.P., and Friis-Hansen, P., Risk modelling of a hydrogen refuelling station using Bayesian network, *International Journal of Hydrogen Energy*, 36 (3), 2389–2397, 2011.
45. MacIntyre, I., Tchouvelev, A.V., Hay, D.R., Wong, J., Grant, J., and Benard, P., Canadian hydrogen safety program, *International Journal of Hydrogen Energy*, 32 (13), 2134–2143, 2007.
46. Middha, P., and Hansen, O.R., Using computational fluid dynamics as a tool for hydrogen safety studies, *Journal of Loss Prevention in the Process Industries*, 22 (3), 295–302, 2009.
47. Middha, P., Hansen, O.R., and Storvik, I.E., Validation of CFD-model for hydrogen dispersion, *Journal of Loss Prevention in the Process Industries*, 22 (6), 1034–1038, 2009.
48. Makarov, D., Verbecke, F., Molkov, V., Roe, O., Skotenne, M., Kotchourko, A., Lelyakin, A., Yanez, J., Hansen, O., Middha, P., Ledin, S., Baraldi, D., Heitsch, M., Efimenko, A., and Gavrikov, A., An inter-comparison exercise on CFD model capabilities to predict a hydrogen explosion in a simulated vehicle refuelling environment, *International Journal of Hydrogen Energy*, 34 (6), 2800–2814, 2009.

49. Middha, P., and Hansen, O.R., CFD simulation study to investigate the risk from hydrogen vehicles in tunnels, *International Journal of Hydrogen Energy*, 34 (14), 5875–5886, 2009.
50. Venetsanos, A.G., Papanikolaou, E., Delichatsios, M., Garcis, J., Hansen, O.R., Heitsch, M., Huser, A., Jahn, W., Jordan, T., Lacome, J.-M., Ledin, H.S., Makarov, D., Middha, P., Studer, E., Tchouvelev, A.V., Teodorczyk, A., Verbecke, F., and Van der Voort, M.M., An inter-comparison exercise on the capabilities of CFD models to predict the short and long term distribution and mixing of hydrogen in a garage, *International Journal of Hydrogen Energy*, 34 (14), 5912–5923, 2009.
51. Baraldi, D., Kotchourko, A., Lelyakin, A., Yanez, J., Middha, P., Hansen, O.R., Gavrikov, A., Efimenko, A., Verbecke, F., Makarov, D., and Molkov, V., An inter-comparison exercise on CFD model capabilities to simulate hydrogen deflagrations in a tunnel, *International Journal of Hydrogen Energy*, 34 (18), 7862–7872, 2009.
52. Venetsanos, A.G., Papanikolaou, E., Hansen, O.R., Middha, P., Garcia, J., Heitsch, M., Baraldi, D., and Adams, P., HySafe standard benchmark problem SBEP-V11: Predictions of hydrogen release and dispersion from a CGH2 bus in an underpass, *International Journal of Hydrogen Energy*, 35 (8), 3857–3867, 2010.
53. Garcia, J., Baraldi, D., Gallego, E., Beccantini, A., Crespo, A., Hansen, O.R., Hoiset, S., Kotchourko, A., Makarov, D., Migoya, E., Molkov, V., Voort, M.M., and Yanez, J., An intercomparison exercise on the capabilities of CFD models to reproduce a large-scale hydrogen deflagration in open atmosphere, *International Journal of Hydrogen Energy*, 35 (9), 4435–4444, 2010.
54. Ham, K., Marangon, A., Middha, P., Versloot, N., Rosmuller, N., Carcassi, M., Hansen, O.R., Schiavetti, M., Papanikolaou, E., Venetsanos, A., Engebo, A., Saw, J.L. Saffers, J.-B., Flores, A., and Serbanescu, D., Benchmark exercise on risk assessment methods applied to a virtual hydrogen refuelling station, *International Journal of Hydrogen Energy*, 36 (3), 2666–2677. 2011.
55. Venetsanos, A.G., Adams, P., Azkarate, I., Bengaouer, A., Brett, L., Carcassi, M.N., Engebo, A., Gallego, E., Gavrikov, A.I., Hansen, O.R., Hawksworth, S., Jordan, T., Kessler, A., Kumar, S., Molkov, V., Nilsen, S., Reinecke, E., Stocklin, M., Schnidtchen, U., Teodorczyk, A., Tigreat, D., and Versloot, N.H.A., On the use of hydrogen in confined spaces: Results from the internal project InsHyde, *International Journal of Hydrogen Energy*, 36 (3), 2693–2699, 2011.
56. Hansen, M.D., and Gammel, G.W., Management of change: A key to safety—not just process safety, *Professional Safety*, 53 (10), 41–50, 2008.
57. EPA, *Catastrophic Failure of Storage Tanks*, Chemical Safety Alert, U.S. Environmental Protection Agency, Washington, D.C., 1997.
58. Alsheyab, M., Jiang, J.-Q., and Stanford, C., Risk assessment of hydrogen gas production in the laboratory scale electrochemical generation of ferrate (VI), *Journal of Chemical Health & Safety*, 15 (5), 16–20, 2008.
59. Xu, P., Zheng, J., Liu, P., Chen, R., Kai, F., and Li, L., Risk identification and control of stationary high-pressure hydrogen storage vessels, *Journal of Loss Prevention in the Process Industries*, 22 (6), 950–953, 2009.
60. *TEEX Developing Hydrogen Fuel Training Program, Occupational Health & Safety* (April 24, 2011).
61. DiBerardinis, L.J. (Editor), *Handbook of Occupational Safety and Health*, 2nd edition, John Wiley & Sons, New York, 1999.

62. Felder, R.M., and Brent, R., Objectively speaking, *Chemical Engineering Education*, 31 (3), 178–179, 1997.
63. Goraya, A., Amyotte, P.R., and Khan, F.I., An inherent safety-based incident investigation methodology, *Process Safety Progress*, 23 (3), 197–205, 2004.
64. Frank, J., Observations on pre-emergency planning, *Industrial Fire World* (June 2007).
65. Hendershot, D.C., Process safety culture, *Journal of Chemical Health and Safety*, 14 (3), 39–40, 2007.
66. Erickson, J.A., Corporate culture. Examining its effects on safety performance, *Professional Safety*, 53 (11), 35–38, 2008.
67. Antonsen, S., The relationship between safety culture and safety on offshore supply vessels, *Safety Science*, 47 (8), 1118–1128, 2009.
68. Hopkins, A., *Thinking about Process Safety Indicators*, Working Paper 53, National Research Centre for OHS Regulation, Australian National University (paper prepared for presentation at the Oil and Gas Industry Conference, Manchester, UK, November 2007).
69. Glendon, I., Safety culture and safety climate: How far have we come and where should we be heading? *Journal of Occupational Health and Safety – Australia and New Zealand*, 24 (3), 249–271, 2008.
70. Amyotte, P.R., and Eckhoff, R.K., Dust explosion causation, prevention and mitigation: An overview, *Journal of Chemical Health and Safety*, 17 (1), 15–28, 2010.
71. Morozzo, C., *Account of a Violent Explosion Which Happened in the Flour-Warehouse, at Turin, December the 14th, 1785; To Which Are Added Some Observations on Spontaneous Inflammations*, From the Memoirs of the Academy of Sciences of Turin (London: The Repertory of Arts and Manufactures, 1795).
72. Amyotte, P.R., *Are Classical Process Safety Concepts Relevant to Nanotechnology Applications?* Nanosafe2010: International Conference on Safe Production and Use of Nanomaterials, *Journal of Physics: Conference Series*, 304, 2011.
73. Hopkins, A. (Editor), *Learning from High Reliability Organisations*, CCH Australia Limited, Sydney, Australia, 2009.

9

HySafe: Safety of Hydrogen as an Energy Carrier

This chapter provides a brief look at one of the more comprehensive hydrogen safety initiatives undertaken worldwide, *HySafe* (Safety of Hydrogen as an Energy Carrier). The intent here is not to give an exhaustive accounting of all HySafe activities and accomplishments, but rather to encourage readers to explore further the various research and educational packages that may be of interest. The chapter is therefore arranged in the following manner: an overview of the Network of Excellence HySafe, followed by a description of important research packages and projects, and an educational curriculum for teaching hydrogen safety.

Two primary resources used in writing this chapter (and hence recommended for further reading) are the HySafe web site [1] and the recent (2011) article by Jordan et al. [2]. Jordan [3] and HySafe [4] provide perspectives on the network as of 2006 and 2007, respectively. Molkov [5], in his preface to the hydrogen safety special issue of the *Journal of Loss Prevention in the Process Industries* published in 2008, provides additional commentary on HySafe activities designed to contribute to safer use of hydrogen and fuel cell technologies.

In his presentation to the 1st European Hydrogen Energy Conference in 2003 (approximately six months before the formal start of the HySafe network), Dorofeev [6] identified the following HySafe objectives:

- To contribute to common understanding and approaches for addressing hydrogen safety issues
- To integrate experience and knowledge on hydrogen safety in Europe
- To integrate and harmonize the fragmented research base
- To provide contributions to EU (European Union) safety requirements, standards and codes of practice
- To contribute to an improved technical culture on handling hydrogen as an energy carrier
- To promote public acceptance of hydrogen technologies

Now in 2011, these same objectives have been concisely expressed by the following [2]:

- To strengthen, focus and integrate the fragmented research on hydrogen safety
- To form a self-sustained competitive scientific and industrial community
- To promote public awareness and trust in hydrogen technologies
- To develop an excellent safety culture

The reason behind presenting these two listings of the same objectives is to demonstrate the evolution of hydrogen safety thinking, even over this relatively short eight-year period. This does not constitute a redefinement of the HySafe objectives, but rather their re-expression in a manner commensurate with recent developments in the safety field. One clearly sees an emphasis in the 2011 formulation of objectives [2] on issues of public *awareness and trust* (compared with *acceptance*) and an excellent *safety culture* (compared with an improved *technical culture*). These are certainly welcome developments that are consistent with arguments previously advanced in this book (e.g., Section 8.3).

9.1 Overview of EC Network of Excellence for Hydrogen Safety

With the above objectives in mind, the Network of Excellence (NoE) HySafe (Safety of Hydrogen as an Energy Carrier) was established and supported by the European Commission (EC), with a start date of March 1, 2004 [2]. A follow-up organization, the International Association for Hydrogen Safety, HySafe (IA), was founded by most of the HySafe consortium members on February 26, 2009 [2]. Official logos for both HySafe and HySafe (IA) are given in Figure 9.1.

The HySafe consortium itself was coordinated by the Forschungszentrum Karlsruhle in Germany [2], now merged with Universitat Karlsruhle since October 1, 2009 to form the Karlsruhle Institute of Technology (KIT) [7]. HySafe participants included approximately 120 scientific researchers from 25 institutions located in 13 countries (12 European countries and Canada) [2]. The organizations involved consisted of 12 public research institutions,

FIGURE 9.1
Logos for Network of Excellence HySafe and the International Association for Hydrogen Safety, HySafe (IA). (From Jordan, T. et al., *International Journal of Hydrogen Energy*, 36 (3), 2656–2665, 2011. With permission.)

TABLE 9.1

Network of Excellence HySafe Consortium Members

Organization	Abbreviation	Country
Forschungszentrum Karlsruhle GmbH	FZK	Germany
L'Air Liquide	AL	France
Federal Institute for Materials Research and Testing	BAM	Germany
BMW Forschung und Technik GmbH	BMW	Germany
Building Research Establishment Ltd.	BRE	United Kingdom
Commissariat a l'Energie Atomique	CEA	France
Det Norske Veritas AS	DNV	Norway
Fraunhofer-Gesellschaft ICT	Fh-ICT	Germany
Forschungszentrum Julich GmbH	FZJ	Germany
GexCon AS	GexCon	Norway
The United Kingdom's Health and Safety Laboratory	HSE/HSL	United Kingdom
Foundation INASMET	INASMET	Spain
Institut National de l'Environnement Industriel et des Risques	INERIS	France
European Commission – JRC – Institute for Energy	JRC	Netherlands
National Center for Scientific Research Demokritos	NCSRD	Greece
StatoilHydro ASA	SH	Norway
DTU/Riso National Laboratory	DTU/Riso	Denmark
TNO	TNO	Netherlands
University of Calgary	UC	Canada
University of Pisa	UNIPI	Italy
Universidad Politecnica de Madrid	UPM	Spain
University of Ulster	UU	United Kingdom
Volvo Technology Corporation	Volvo	Sweden
Warsaw University of Technology	WUT	Poland
Russian Research Centre Kurchatov Institiute	KI	Russia

Source: From Jordan, T. et al., *International Journal of Hydrogen Energy*, 36 (3), 2656–2665, 2011. With permission.

seven industrial partners, five universities, and one governmental body [2]; details are given in Table 9.1. Further thoughts on the origin and objectives of the HySafe network have been given by Molkov [8].

9.2 HySafe Work Packages and Projects

As described (and illustrated graphically) by Jordan et al. [2], HySafe activities consisted of 15 work packages and three internal projects accommodated within four "activity clusters":

- Basic research
- Risk management
- Dissemination
- Management

The above clusters indicate a primary focus on knowledge acquisition through fundamental and applied research, coupled with effective overall management and publication and presentation of findings.

The 15 work packages as given on the HySafe web site [1] are as follows:

- Biennial report on hydrogen safety
- Integration of experimental facilities
- Scenario and phenomena ranking
- Hydrogen incident and accident database
- Principal CFD (computational fluid dynamics) exercises and guidelines
- Mapping priorities and assessment
- Hydrogen release, distribution, and mixing
- Hydrogen ignition and jet fires
- Hydrogen explosions
- Mitigation
- Risk assessment methodologies
- International conference on hydrogen safety
- e-Academy of hydrogen safety (discussed in Section 9.3)
- Contribution to standards and legal requirements
- Material compatibility and structural integrity

Not surprisingly, one sees in the above work packages a clear linkage to the activity clusters focused on basic and applied research, network management, and dissemination of both research and educational products.

Also identified on the HySafe web site [1] are the three internal projects:

- InsHyde
    - This project had its origins in the understanding that hydrogen releases in confined and partially confined environments—even seemingly small releases with low flow rates—could pose significant problems because of the potential for ignition of flammable gas accumulations leading to deflagration and possibly detonation [2]. Issues related to the indoor use of hydrogen were investigated under the InsHyde project by means of a combination of theoretical and experimental studies on the release, mixing,

    and combustion of hydrogen [2]. Sensor evaluations were also conducted [2], thus addressing the active engineered category of the hierarchy of safety controls (Section 7.1).

- HyTunnel
    - The HyTunnel project also dealt with confined environments—as the name itself illustrates, the specific geometry being a tunnel. Motivation for this work came from the need for management of the hazards and risks related to tunnel fires, particularly in light of the relevant European Union regulations concerning road-tunnel safety [2]. In addition to the performance of fire and explosion simulations, numerical modeling was used extensively in an attempt to better understand the complex interaction among the following parameters affecting dispersion: (1) hydrogen release, (2) tunnel geometry, and (3) ventilation system design [2]. Buoyancy, one of the inherent safety aspects of hydrogen discussed in Chapter 7, was observed to play a key role in these dispersion studies wherein increased tunnel height yielded a lower risk profile [2].
- HyQRA
    - This project served as a vehicle for linking fundamental scientific research to industrially relevant applications [2]. The primary HyQRA objective, again as the name implies, was to develop quantitative risk assessment (QRA) tools for hydrogen technologies that were appropriate for the level of detail required and/or feasible [2]. Improvements were sought for various aspects of the QRA process: (1) screening models, (2) scenario selection, (3) ignition probability, (4) fire modeling, (5) structural response, and (6) acceptance criteria [2]. Several of these points were previously discussed in Section 8.2.4 under the PSM element *process risk management*.

The Hysafe web site [1] and Jordan et al. [2] provide information on the products (i.e., deliverables) arising from the various work packages and internal projects listed above. Several research findings have also been published in the archival literature; selected illustrative examples from the *International Journal of Hydrogen Energy* are now presented to close the current section. The emphasis in these examples is clearly on both process safety and occupational safety (as defined and discussed in Section 8.1), although not necessarily in what might be termed a "pure" process environment such as the hydrogen upgrader section of an oil refinery. Industrial workers must obviously be protected from hydrogen hazards but so too must the public. Hence one sees in these HySafe deliverables a focus on application to garages, transportation tunnels, and refueling stations.

One of the rigorous scientific research features of the HySafe work was the use of standard benchmark exercise problems (SBEPs). Venetsanos et al. [9] provide details on one of the InsHyde SBEPs aimed at predicting distribution and mixing patterns of hydrogen in a garage. Ten different CFD codes employing eight different turbulence submodels were used in this intercomparison exercise in an attempt to reduce the variation in predicted results from the numerical simulators [9]. In a subsequent publication, Venetsanos et al. [10] summarize the results from the InsHyde project in terms of, among other features, experimental data obtained from dispersion and combustion experiments, performance evaluation of commercial hydrogen detectors, and further CFD code intercomparisons.

In a similar vein, Baraldi et al. [11] describe a HyTunnel SBEP in which five CFD codes, again having different turbulence and combustion submodels, were used to simulate explosions of stoichiometric hydrogen-air mixtures in a 78.5-m long tunnel. The numerical simulation results were compared against one another and also with experimental data obtained for the same geometry; good agreement was determined to exist with respect to experimental and predicted maximum overpressures [11]. Middha and Hansen [12] used the GexCon code FLACS (Section 8.2.4) to conduct hydrogen-vehicle risk assessments for different tunnel configurations and both worst-case and more realistic (i.e., credible) scenarios. Venetsanos et al. [13] investigated, again using various CFD codes, hydrogen release and dispersion from a hydrogen-fueled bus in an underpass; that is, a semiconfined environment as in the previously referenced tunnel studies, but also one having roof obstructions as would be present in roof-and-slab style construction.

There are two fundamental aspects to QRA: quantification of probability (or likelihood) and quantification of consequence severity. Garcia et al. [14] conducted an intercomparison exercise using CFD codes to predict the latter of these two risk parameters—in this case for the explosion of a 2094-m^3 stoichiometric hydrogen-air mixture in an unconfined environment. The practical application of this work was envisaged as an accidental leak and subsequent explosion of hydrogen at a facility such as a refueling station [14]. This scenario was considered by Ham et al. [15] in their application of various QRA methodologies involving both analytical and CFD consequence modeling to a "virtual" hydrogen refueling station complete with vulnerable surroundings. A key finding of this research was that the numerical models provided more realistic results in the near-field and/or confined areas, while there was no discernable advantage of numerical over analytical processing in the open air [15]. Clearly there is a need for further work in this area which has implications for reducing risk to both on-site workers and off-site members of the general public [15].

9.3 e-Academy of Hydrogen Safety

The previous section has highlighted some of the research achievements of the HySafe Network of Excellence—in particular the issues of CFD code validation with experimental data and the assessment of underlying code assumptions and submodels. To close this chapter we turn to the matter of education and training in hydrogen safety.

Dahoe and Molkov [16] describe the establishment of an "e-academy of hydrogen safety" led by the University of Ulster in collaboration with other HySafe partners. They give details on a postgraduate certificate course in hydrogen safety delivered in the distance-learning mode, as well as an annual European summer school in hydrogen safety. Both of these offerings are aimed at helping to achieve the following objectives of the e-academy of hydrogen safety [1]:

- Integration of academic and other institutions through the development and implementation of an international curriculum on hydrogen safety engineering.
- Development of a database of organizations working in the hydrogen industry to form a market of potential trainees and to disseminate results of mutual activities of the network.
- Creation of a pool of specialists for the delivery of joint training/teaching on hydrogen safety and joint supervision of research students.

The key element in the first of the above objectives is the international curriculum on hydrogen safety engineering (the "backbone" of the e-academy [16]). The current version of this curriculum posted on the HySafe web site [1] is shown in Table 9.2. Topics are arranged in 15 modules organized into three thematic areas—basic, fundamental, and applied—with the majority of the basic modules intended for the undergraduate level and the majority of the fundamental and applied modules designed for the graduate level [1]. This thinking is consistent with the graphic displayed in Figure 9.2, which underpins the developers' philosophy of the undergraduate program being grounded with a strong engineering science core and augmented with educational material focused on hydrogen safety [1]. The graduate program then covers more advanced hydrogen safety topics consistent with the overall HySafe research activities [1].

What one sees in this approach is a deliberate attempt to integrate hydrogen safety material into the undergraduate curriculum while providing opportunities for acquisition of specialized knowledge at the graduate (e.g., master's) level. This is an approach the current authors wholeheartedly endorse and which is consistent with arguments previously advanced for the topic of risk management [17]. Further, the international curriculum shown in Table 9.2 provides numerous examples of possible training topics on

TABLE 9.2

Structure of the International Curriculum on Hydrogen Safety Engineering

Level	Module	Contents
Basic	Thermodynamics	Fundamental concepts and first principles; volumetric properties of a pure substance; first law of thermodynamics; first law of thermodynamics and flow processes; second law of thermodynamics; second law of thermodynamics and flow processes; first and second laws of thermodynamics, and chemically reacting systems; phase equilibrium; thermodynamics and electrochemistry
	Chemical kinetics	Rates of chemical reactions; kinetics of complex reactions; surface reactions; application of sensitivity analysis to reaction mechanisms; reduction of complex reaction systems to simpler reaction mechanisms; chemical kinetics and the detonation of combustible mixtures
	Fluid dynamics	Fluid statics; kinematics of the flow field; kinematics of incompressible potential flow; kinematics of compressible flow; incompressible laminar viscous flow; mathematical models of fluid motion; dimensional analysis and similitude; incompressible turbulent flow; waves in fluids and the stability of fluid flow; compressible turbulent flow
	Heat and mass transfer	Basic modes of heat transfer and particular laws; isothermal mass transfer; heat conduction; convection heat transfer; forced convection; natural convection; heat transfer with phase change; radiation heat transfer; simultaneous heat and mass transfer
	Solid mechanics	Analysis of stress; deformation and strain; tension and compression; statically indeterminate force systems; thin-walled pressure vessels; direct shear stresses; torsion; shearing force and bending moment; centroids, moments of inertia, and products of inertia of plane areas; stresses in beams; elastic deformation of beams: double integration method; statically indeterminate elastic beams; special topics in elastic beam theory; plastic deformation of beams; columns; strain energy methods; combined stresses; members subject to combined loadings
Fundamental	Hydrogen as an energy carrier	Introduction to hydrogen as an energy carrier; introduction to hydrogen applications and case studies; equipment for hydrogen applications; possible accident scenarios; definitions and overview of phenomena and methodologies related to hydrogen safety

TABLE 9.2 (*Continued*)

Structure of the International Curriculum on Hydrogen Safety Engineering

Level	Module	Contents
	Fundamentals of hydrogen safety	Hydrogen properties; compatibility of metallic materials with hydrogen; hydrogen thermochemistry; governing equations of multicomponent reacting flows; premixed flames; diffusion flames; partially premixed flames; turbulent premixed combustion; turbulent non-premixed combustion; ignition and burning of liquids and solids; fire through porous media
	Release, mixing, and distribution	Fundamentals of hydrogen release and mixing; handling hydrogen releases
	Hydrogen ignition	Hydrogen ignition properties and ignition sources; prevention of hydrogen ignition
	Hydrogen fires	Fundamentals of hydrogen fires
	Explosions: deflagrations and detonations	Deflagrations; detonation; transitional hydrogen explosion phenomena
Applied	Fire and explosion effects on people, structures, and the environment	Thermal effects of hydrogen combustion; blast waves; calculation of pressure effects of explosions; structural response, fragmentation and missile effects; fracture mechanics
	Accident prevention and mitigation	Prevention, protection and mitigation; basic phenomena underpinning mitigation technologies; standards, regulations, and good practices related to hydrogen safety; inertization; containment; explosion venting; flame arresters and detonation arresters
	Computational hydrogen safety engineering	Introduction to CFD; introduction to thermodynamic and kinetic modeling; mathematical models in fluid dynamics; finite difference method; solution of the generic transport equation; solution of weakly compressible Navier-Stokes equations; solution of compressible Navier-Stokes equations; turbulent flow modeling; combustion modeling; high-speed reactive flows; modeling of hydrogen-air diffusion flames and turbulence-radiation interactions; modeling of liquid hydrogen pool fires; multiphase flows; special topics
	Risk assessment	General risk assessment and protective measures for hazardous materials processing and handling; regulations, codes, and standards; risk assessment methodologies; hazard identification and scenario development; effect analysis of hydrogen accidents; vulnerability analysis; risk reduction and control in the hydrogen economy

Source: From HySafe web site, http://www.hysafe.net/.

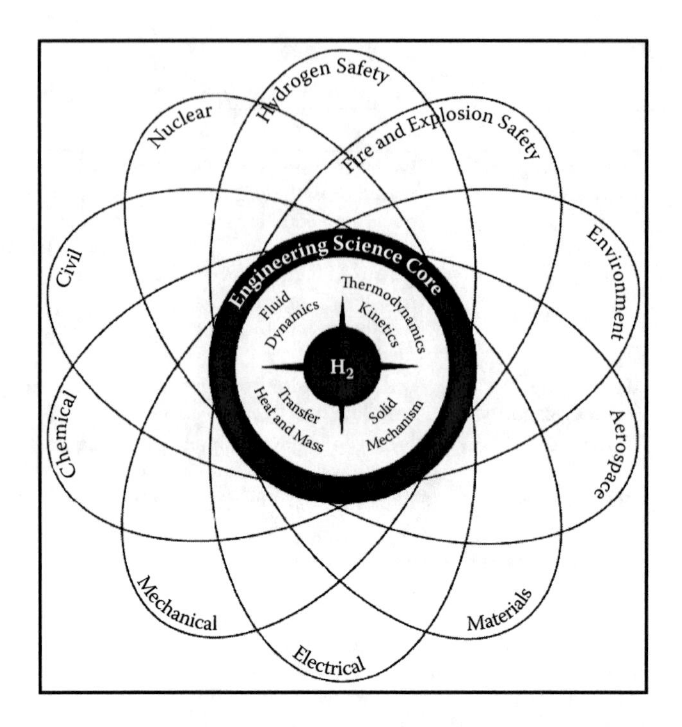

FIGURE 9.2
Hydrogen safety in relation to other branches of engineering science (from HySafe [1]).

hydrogen safety according to the hierarchical learning continuum known as Bloom's Taxonomy of Educational Objectives (described in Section 8.2.8). The fundamental and applied modular topics, although more extensive in depth of coverage, are also generally consistent with the scope of the current book.

In keeping with the introductory remarks to this chapter, our intention here has been to highlight key research and educational achievements of the HySafe Network of Excellence. The HySafe web site [1] and the summary paper by Jordan et al. [2] are recommended as starting points for further study.

References

1. *HySafe – Safety of Hydrogen as an Energy Carrier*, http://www.hysafe.net/ (accessed August 4, 2011).
2. Jordan, T., Adams, P., Azkarate, I., Baraldi, D., Barthelemy, H., Bauwens, L., Bengaouer, A., Brennan, S., Carcassi, M., Dahoe, A., Eisenrich, N., Engebo, A., Funnemark, E., Gallego, E., Gavrikov, A., Haland, E., Hansen, A.M., Haugom, G.P., Hawksworth, S., Jedicke, O., Kessler, A., Kotchourko, A., Kumar, S., Langer, G., Stefan, L., Lelyakin, A., Makarov, D., Marangon, A., Markert, F.,

Middha, P., Molkov, V., Nilsen, S., Papanikolaou, E., Perrette, L., Reinecke, E.-A., Schmidtchen, U., Serre-Combe, P., Stocklin, M., Sully, A., Teodorczyk, A., Tigreat, D., Venetsanos, A., Verfondern, K., Versloot, N., Vetere, A., Wilms, M., and Zaretskiy, N., Achievements of the EC Network of Excellence HySafe, *International Journal of Hydrogen Energy*, 36 (3), 2656–2665, 2011.

3. Jordan, T., *HySafe – The Network of Excellence for Hydrogen Safety*, WHEC 16, Lyon, France (June 13–16, 2006).
4. HySafe, 2004 – 2009: A Network of Excellence in the 6th Framework Programme of the European Commission, with 25 European Partners, HYSAFE – Safe Use of Hydrogen as an Energy Carrier (2007).
5. Molkov, V., Preface, Journal of Loss Prevention in the Process Industries, Special Issue on Hydrogen Safety, 21 (2), 129–130, 2008.
6. Dorofeev, S., Safety aspects of hydrogen as an energy carrier, Presented at 1st European Hydrogen Energy Conference, Grenoble, France (September 2–5, 2003).
7. KIT – Karlsruhle Institute of Technology, http://www.kit.edu/kit/english/index.php (accessed August 5, 2011).
8. Molkov, V., Hydrogen safety research: State-of-the-art, in *Proceedings of the 5th International Seminar on Fire and Explosion Hazards*, Edinburgh, UK (April 23–27, 2007).
9. Venetsanos, A.G., Papanikolaou, E., Delichatsios, M., Garcis, J., Hansen, O.R., Heitsch, M., Huser, A., Jahn, W., Jordan, T., Lacome, J.-M., Ledin, H.S., Makarov, D., Middha, P., Studer, E., Tchouvelev, A.V., Teodorczyk, A., Verbecke, F., and Van der Voort, M.M., An inter-comparison exercise on the capabilities of CFD models to predict the short and long term distribution and mixing of hydrogen in a garage, *International Journal of Hydrogen Energy*, 34 (14), 5912–5923, 2009.
10. Venetsanos, A.G., Adams, P., Azkarate, I., Bengaouer, A., Brett, L., Carcassi, M.N., Engebo, A., Gallego, E., Gavrikov, A.I., Hansen, O.R., Hawksworth, S., Jordan, T., Kessler, A., Kumar, S., Molkov, V., Nilsen, S., Reinecke, E., Stocklin, M., Schnidtchen, U., Teodorczyk, A., Tigreat, D., and Versloot, N.H.A., On the use of hydrogen in confined spaces: Results from the internal project InsHyde, *International Journal of Hydrogen Energy*, 36 (3), 2693–2699, 2011.
11. Baraldi, D., Kotchourko, A., Lelyakin, A., Yanez, J., Middha, P., Hansen, O.R., Gavrikov, A., Efimenko, A., Verbecke, F., Makarov, D., and Molkov, V., An inter-comparison exercise on CFD model capabilities to simulate hydrogen deflagrations in a tunnel, *International Journal of Hydrogen Energy*, 34 (18), 7862–7872, 2009.
12. Middha, P., and Hansen, O.R., CFD simulation study to investigate the risk from hydrogen vehicles in tunnels, *International Journal of Hydrogen Energy*, 34 (14), 5875–5886, 2009.
13. Venetsanos, A.G., Papanikolaou, E., Hansen, O.R., Middha, P., Garcia, J., Heitsch, M., Baraldi, D., and Adams, P., HySafe standard benchmark problem SBEP-V11: Predictions of hydrogen release and dispersion from a CGH2 bus in an underpass, *International Journal of Hydrogen Energy*, 35 (8), 3857–3867, 2010.
14. Garcia, J., Baraldi, D., Gallego, E., Beccantini, A., Crespo, A., Hansen, O.R., Hoiset, S., Kotchourko, A., Makarov, D., Migoya, E., Molkov, V., Voort, M.M., and Yanez, J., An intercomparison exercise on the capabilities of CFD models to reproduce a large-scale hydrogen deflagration in open atmosphere, *International Journal of Hydrogen Energy*, 35 (9), 4435–4444, 2010.

15. Ham, K., Marangon, A., Middha, P., Versloot, N., Rosmuller, N., Carcassi, M., Hansen, O.R., Schiavetti, M., Papanikolaou, E., Venetsanos, A., Engebo, A., Saw, J.L. Saffers, J.-B., Flores, A., and Serbanescu, D., Benchmark exercise on risk assessment methods applied to a virtual hydrogen refuelling station, *International Journal of Hydrogen Energy*, 36 (3), 2666–2677, 2011.
16. Dahoe, A.E., and Molkov, V.V., On the implementation of an international curriculum on hydrogen safety engineering into higher education, *Journal of Loss Prevention in the Process Industries*, 21 (2), 222–224, 2008.
17. Amyotte, P.R., and McCutcheon, D.J., *Risk Management: An Area of Knowledge for all Engineers*, Discussion paper prepared for Canadian Council of Professional Engineers, Ottawa, Ontario, 2006.

10

Case Studies

The purpose of this chapter is to present guidance on the development and use of case studies (or as they are sometimes called, case histories). The role of incident investigation reports in this regard is highlighted and several examples drawn from a range of industrial applications involving hydrogen are given.

To answer the question of whether case studies are helpful in enhancing the overall safety initiative, one need only consider the following quote from Crowl and Louvar [1]: "To paraphrase G. Santayana, one learns from history or one is doomed to repeat it." Case studies therefore incorporate the concept of *lessons learned* as the primary motivation for their use. This usually means that the most effective case studies are those giving details of a failure or shortcoming of some sort; it is human nature to pay attention when a story is being told and a loss—whether catastrophic or not—is involved [2]. The above quote illustrates the importance of learning from case histories and avoiding hazardous situations; the alternative is to ignore the mistakes of others and be involved in potentially life-threatening incidents [1, 2].

10.1 Case Study Development

The development of a case study involves first the selection of an appropriate incident (accident or near-miss); this is followed by an exposition of the lessons that can be learned from the incident and that ideally have universal or at least industry-specific application. For example, Chapter 2 describes several incidents and lessons learned that are drawn from the archival literature and various databases, the latter relying on incident investigation reports as discussed here in Section 10.3.

With respect to inherently safer design (ISD as presented in Chapter 7), Kletz and Amyotte [2] have presented thoughts on the following 12 avenues for developing case studies:

- Everyday life experiences
- Newspapers and magazines
- Topical conferences
- Technical papers

- Process safety books
- Case study books
- Other inherent safety books
- Books from other industries and applications
- Training packages
- Trade literature
- *Loss Prevention Bulletin*
- Chemical Safety Board reports

Because ISD is concept based, some of the suggestions above may not translate literally to the world of hydrogen safety, which is focused on the nature and use of a particular substance. Few people would have first-hand experience in their daily lives with hydrogen, while many more would be able to relate to the notion that a tricycle (having three wheels) is inherently safer (more stable) than a bicycle (having two wheels).

The popular media (newspapers and magazines), on the other hand, seem to be a ready source of hydrogen-related case information, at least with respect to incident descriptions if not lessons learned beyond the impact of such incidents on public perception. Witness the numerous articles published in the *New York Times* on, for example, hydrogen leaking from a road tanker involved in an accident while carrying 11,000 kg of compressed hydrogen [3], liquid hydrogen fuel leaks during operation of the U.S. space shuttle program [4–6], and gaseous hydrogen leaks from the non-nuclear operations (electrical generator cooling system) of a nuclear power plant [7].

Topical conferences such as the International Conference on Hydrogen Safety (ICHS) may prove to be quite helpful with respect to case study development. The fourth edition of the ICHS (September 12–14, 2011, in San Francisco, California) encompasses the topic of case studies under the theme of risk management of hydrogen technologies [8].

Similarly, technical papers cited throughout this book have a key role to play in developing hydrogen safety case studies. General papers such as Sanders [9] contain information on lessons related to items including human error and human factors, which can be easily tailored to applications involving hydrogen. Hydrogen-specific technical case study papers abound in the archival literature (see Chapter 2). As an additional example also referenced later in Section 10.3, Weiner, Fassbender, and Quick [10] present an interesting combination of *Hydrogen Safety Best Practices* (www.h2bestpractices .org) and *Hydrogen Incident Reporting and Lessons Learned* (www.h2incidents.org).

Process safety books (e.g., Crowl and Louvar [1]) and case study books (e.g., Kletz [11]) provide both general cases and widely applicable lessons, as well as specific information on hydrogen-related incidents such as the 1989 Pasadena, Texas, vapor cloud explosion (see Chapters 2 and 8). There are, of course other books on hydrogen usage and safety (e.g., Gupta [12]); these may also provide relevant case study details. Books from other industries and applications would seem at first glance to be of limited use in hydrogen case study development. These texts have proven, however, to be quite useful in demonstrating the general lessons on safety culture (Section 8.3) to be learned from the railway and aircraft maintenance industries [13] and air traffic control activities [14].

Even only a quick Internet search for hydrogen training packages yields multiple hits (for example, reference [15]). Some Internet sources provide helpful case studies on topics such as hydrogen-induced crack-resistant (HIC) steel [16] and hydrogen fueling stations [17].

Trade (or industry sector) literature sources often provide incident descriptions that can be followed up to develop case studies, such as the explosion and fire at an ammonia plant involving a mixture of gases including hydrogen, nitrogen, and ammonia [18]. The same publication also describes problems associated with the degradation of munitions in stockpiles maintained by the U.S. army [19]. One potential issue is the explosion risk from hydrogen forming as stockpiled mustard gas degrades; risk reduction recommendations given include regular testing of all stored chemical agents, establishing databases for accurate information recording, and using statistical analysis for early identification of trends [19]. Interestingly, the inherently safer approach of swiftly destroying all the munitions is described as "ultimately the only effective way to reduce risks to the public" [19].

From the world of occupational health and safety comes a general (non-incident-specific) case study on battery fire and explosion hazards [20]. In the final stages of battery recharging, hydrogen can be produced by a process known as *gassing*; recommended risk reduction measures from the hierarchy of controls include air movement and exchange, hydrogen gas monitoring, and appropriate signage aimed at avoiding ignition sources [20]. In accordance with the discussion in Section 10.3, post-incident investigations of battery fires and explosions often indicate causes related to operational errors and safety shortcuts [20]. To further develop the lessons learned from these incidents, one must ask whether root causes at the management system level existed (as per Sections 10.2.3 and 10.3), and whether inherently safer prevention and mitigation measures were available.

The *Loss Prevention Bulletin* (LPB) is published by the Institution of Chemical Engineers (IChemE) in the United Kingdom. LPB issues represent a good source of case studies arising from process accident and near-miss reports [2]. For example, Carson and Mumford [21] illustrate various concerns related to the hazards and control of batch processing. In a listing of potentially hazardous exothermic processes, they include both *hydrogenation* (addition of hydrogen atoms to both sides of a double- or triple-bond) and *halogenation* (substitution of atoms such as hydrogen in organic molecules by halogens). In keeping with the case study philosophy of describing incidents and elucidating the accompanying lessons learned, an online search of the

TABLE 10.1

Hydrogen Case Study Examples Found in the *Loss Prevention Bulletin*

Issue	Year	Case Study
015	1977	Fires involving hydrogen in a naphtha cracker and in a hydrogenation reaction
068	1986	Vapor cloud explosions involving hydrogen-rich gases
083	1988	Failure of a woven-steel braided flexible hose on a hydrogen installation area
156	2000	Explosion of hydrogen in a pipeline intended for transfer of CO_2 gas from an ammonia plant
207	2009	Hydrogen explosions from charging batteries; see also Ramsey [20]
		Unexpected generation of hydrogen in a new process for production of aluminum chloride
		Hydrogen generation inside sealed components at a refinery storage terminal
215	2010	Hydrogen explosion in an electrolyter plant

Source: Institute of Chemical Engineers, http://www.icheme.org/sitecore/content/icheme_home/shop/search results.aspx? keywords=hydrogen&product=&CurrentPage=1&SortBy=Relevance&OrderBy=Asce.

Loss Prevention Bulletin with the keyword *hydrogen* will reveal several LPB issues [22] that provide such information. Examples are given in Table 10.1.

As noted on its web site [23], the U.S. Chemical Safety and Hazard Investigation Board (Chemical Safety Board or CSB) is an independent, non-regulatory federal agency that conducts root cause investigations of chemical accidents at fixed industrial facilities. The reports of its investigations are available on the CSB web site [23] for downloading and are often accompanied by video footage and animation of the incident sequence and root causes/lessons learned [2]. The safety bulletin [24, 25] on the ISD practice of *making incorrect assembly impossible* (as previously described in Section 7.6) is an excellent hydrogen-specific example of the significant case study value of CSB reports. This point is reiterated in Section 10.3 and other CSB examples are given in Section 10.4.

In short, there is an abundance of mechanisms for developing case studies relating to the lessons that can be learned from occupational and process incidents involving hydrogen. The next section suggests ways to use these case studies once they have been sourced and/or prepared.

10.2 Use of Case Studies

Case studies are particularly useful in conducting training exercises as part of an overall process safety management (PSM) scheme (see Section 8.2.6).

They have also been shown to have value in addressing other PSM components such as hazard identification and risk analysis [26] (see Table 8.2). Further, educational efforts within undergraduate and graduate science and engineering programs can be significantly enhanced by the use of case studies to emphasize the practical implications of various safety concepts and methodologies.

As discussed in the introduction to the current chapter, it is the concept of *lessons learned* that makes case studies so effective regardless of their ultimate end use. The lessons that can be learned by the study of previous incidents fall within one or more of the following general categories: (1) legacy lessons, (2) engineering lessons, and (3) management lessons. The meaning of each of these categories can be best understood by briefly examining the Bhopal, India, incident—without doubt, the most significant process safety event that has occurred worldwide to the present time.

As excerpted from Khan and Amyotte [27], the Bhopal plant was a pesticide-producing facility owned by the Union Carbide Corporation of the United States and local interests. The plant was located in the town of Bhopal, the capital of Madhya Pradesh state in central India. At 12:45 am on December 3, 1984, several thousand kilograms of a toxic gas, rich in methyl isocyanate (MIC), were released into the atmosphere from the plant, causing over 2000 civilian casualties [28].

The physical properties of MIC make it highly hazardous It has a flash point of –18.1°C and boils at 39.1°C, reacts exothermically with water, and has a maximum allowable 8-h exposure concentration as low as 0.02 ppm. MIC was the intermediate product of carbaryl (a pesticide) manufacturing. It was stored in three horizontal stainless steel vessels, referred to as tanks 610, 611, and 619. Tanks 610 and 611 were used in normal operation and the third tank, 619, was for emergency use. The MIC was usually stored under refrigerated conditions, but in the days before the incident, the refrigeration unit had been shut down.

On the day of the incident, storage tank 610 became contaminated with some other substance (most probably water). An exothermic reaction heated the MIC to a temperature beyond its boiling point. This led to the concrete mounds above the tank cracking and MIC vapor being released through a relief valve. The scrubber and flare systems designed to remove and destroy the MIC were not in operation on the day of the incident. After an hour and a quarter, the relief valve on tank 610 repositioned and the MIC release was stopped. By that time about 36,000 kg of material had escaped from tank 610, of which about 25, 000 kg were MIC vapor [1, 28].

10.2.1 Case Study Legacy Lessons

The legacy of Bhopal is its long-lasting impact on the chemical process industries and numerous other areas of industrial endeavor [29–31]. As described by process safety expert Dennis Hendershot (West et al. [29]): "A significant impact of Bhopal was to make everybody—corporate management, government, communities—aware of the potential magnitude of a chemical accident." The results of this heightened awareness have been manifested positively in new process safety regulatory regimes, worldwide best-practice initiatives such as Responsible Care®, and a greater sense of the need for corporations to export technology—not unacceptable risk—to the developing world.

Most importantly, however, coupled forever with Bhopal are the chemical MIC and the immeasurable loss caused to thousands of human beings in 1984—loss that continues to this day in the form of ongoing suffering from respiratory distress caused by exposure to MIC and other chemicals remaining on the plant site, which have the potential for further health and environmental impacts [30]. As illustrated in Figure 10.1, the legacy of a process incident can involve negative and positive aspects.

Arguably, the closest the hydrogen industry has come to having a case study with Bhopal-type legacy lessons is the *Hindenburg* disaster (and perhaps also the *Challenger* explosion) described in Chapter 2. Whether warranted or not, most people who are aware of the *Hindenburg* will automatically associate that word with another—*hydrogen*. This may not be especially problematic with respect to air travel given that few people will experience a trip on a buoyancy-driven airship during their lifetime. It is imperative, however, that the legacy lessons from incidents such as the *Hindenburg* not be forgotten as public travel via hydrogen-fueled vehicles grows in feasibility and popularity.

FIGURE 10.1
Memorial commemorating Bhopal disaster. (From *Bhopal Memorial*, http://upload.wikimedia.org/wikipedia/commons/d/d8/Bhopal-Union_ Carbide_1.jpg. Accessed September 8, 2011.)

10.2.2 Case Study Engineering Lessons

The engineering lessons of a case study relate to the hierarchy of safety controls as discussed in Section 7.1. With respect to Bhopal, it is evident that safety at the facility was heavily dependent on engineered and procedural safeguards, the effectiveness and operational state of which were questionable. This catastrophe might have been prevented, or the consequences mitigated, through appropriate inherent safety considerations [27].

The relevance of the above paragraph to the hydrogen industry can be found in Chapter 7. Following a specific incident involving hydrogen, the engineering lessons that are transferable to other hydrogen applications would include the use and effectiveness of passive engineered, active engineered, and procedural measures, as well as the potential for implementation of the principles of inherently safer design (ISD). For example, the case of a gaseous hydrogen release necessitates an examination of the detection and mitigation measures in place, and should also be followed by investigation of hydrogen storage possibilities in a liquid state or as metal hydrides.

10.2.3 Case Study Management Lessons

Root cause analysis of an incident must result in identification of the lessons to be learned by management—both the organizational personnel in management positions and the safety management system itself. As described in the previous section, many of the engineered and procedural safeguards at the Bhopal plant were nonfunctional, and there was a lack of commitment to inherently safer design. Where then does responsibility for these shortcomings exist? Squarely on the shoulders of management; no modern root cause analysis methodology could conclude otherwise. (See Section 8.1.)

From a hydrogen safety perspective, the issues of process safety management and safety culture are described in Chapter 8. Deficiencies in these areas constitute key case study lessons that are critical to the prevention of future incidents. This point is expanded upon in the following section.

10.3 Incident Investigation Reports

Management deficiencies identified via incident investigation (company-sponsored or third-party) represent universal findings that can be generalized across industry sectors. Lack of adequate loss prevention and management is generally due to deficiencies in one or more of three areas: the safety management system itself, the standards identified and set for the safety management system, and the degree of compliance with such standards [33]. A particular management system element may have missing

FIGURE 10.2
Westray No. 1 main portal post-explosion [34].

components or may be entirely absent; alternatively, the management system element may be present to some degree, but could have inappropriate standards, or perhaps standards with which there is little or no compliance [33].

These points are (unfortunately) well illustrated in the Westray case study, as demonstrated by the following excerpt from Amyotte and Eckhoff [33].

The Westray coal mine explosion occurred in Plymouth, Nova Scotia, Canada, on May 9, 1992, killing 26 miners. An indication of the destructive overpressures generated underground can be seen in Figure 10.2, which shows surface damage at the mine site. The methane levels in the mine were consistently higher than regulations, which was caused by inadequate ventilation in the mine. Dust accumulations also exceeded permissible levels due to inadequate cleanup of coal dust; additionally, there was no crew in charge of rock dusting (inerting the coal dust with limestone or dolomite). These and many other factors contributed to the poor work conditions that continually existed in the Westray mine and made it the site of an incident waiting to happen. All of these substandard conditions and practices could be attributed to the lack of concern that management had towards safety issues in the mine, which was one of the primary root causes of the problem at Westray.

Management system elements (see Chapter 8) that contributed to the Westray explosion include inadequate:

- Management commitment and accountability to safety matters (which is a key element in establishing an effective company safety culture).
- Management of change procedures.

- Incident investigation (including near-miss reporting and investigation).
- Training (orientation, safety, task-related, etc.).
- Task definition and safe work practices and procedures.
- Workplace inspections and more proactive hazard identification methodologies.
- Program evaluation and audits.

System standards that contributed to the Westray explosion include standards (i.e., levels of performance) relating to virtually all the system elements listed above, including inadequate:

- Concern expressed by management toward safety matters (in terms of the standard of care one would reasonably expect and which, from a legal perspective, should be considered mandatory).
- Follow-through on inspections for substandard practices and conditions.
- Action on hazard reports submitted by employees.
- Job instructions for employees.
- Equipment maintenance.
- Scheduling of management/employee meetings to discuss safety concerns.

Compliance factors that contributed to the Westray explosion include:

- Poor correlation between management actions and official company policy concerning the relationship between safety and production (as evidenced by the same management personnel holding responsibility for both production and underground safety).
- Inadequate compliance with industry practice and legislated standards concerning numerous aspects of coal mining, such as methane concentrations, rock dusting, control of ignition sources underground, etc.

This type of analysis, that is, critical assessment of the management-level root causes of an incident, is routinely conducted by the U.S. Chemical Safety Board (CSB) as described in Section 10.1 for a case study involving hydrogen. Other CSB examples for hydrogen-related incidents are given in Section 10.4 with an emphasis on management-level issues. These additional case study illustrations are consistent with the discussion in Chapter 2 and Section 8.1 concerning the 1989 Pasadena, Texas, vapor cloud explosion in which the management system concepts of hazard assessment and maintenance control were determined to be inadequate.

As also described in Chapter 2, several accident databases have been used to collect and classify past accidents related to hydrogen applications. By definition, such databases are the result of incident descriptions and/or investigation reports, the quality of which can be quite variable (ranging from newspaper articles to public inquiry findings). Database analysis can nevertheless be quite valuable as a learning method; in the words of Sepeda [35]:

> Learning from the experiences of others . . . is an essential tool since industry has neither the time and resources nor the willingness to experience an incident before taking corrective or preventive steps.

Useful databases include the general Major Hazard Incident Database Service (MHIDAS) [36–38], the material-specific Hydrogen Incident Reporting and Lessons Learned (H2Incidents)/Hydrogen Safety Best Practices (H2BestPractices) [10], and the HySafe work package (Chapter 9), Hydrogen Incidents and Accidents Database (HIAD) [39].

Figure 10.3 gives an example of the lessons that can be learned from such databases, in this case, MHIDAS. Here we see the results of the analysis conducted by Gerboni and Salvador [36], who demonstrated the importance of minimizing human error and mechanical failure to prevent accidents involving hydrogen. These findings also illustrate the need for effective process safety management according to the corresponding PSM elements of *human factors* (Section 8.2.7) and *process and equipment integrity* (Section 8.2.6), respectively.

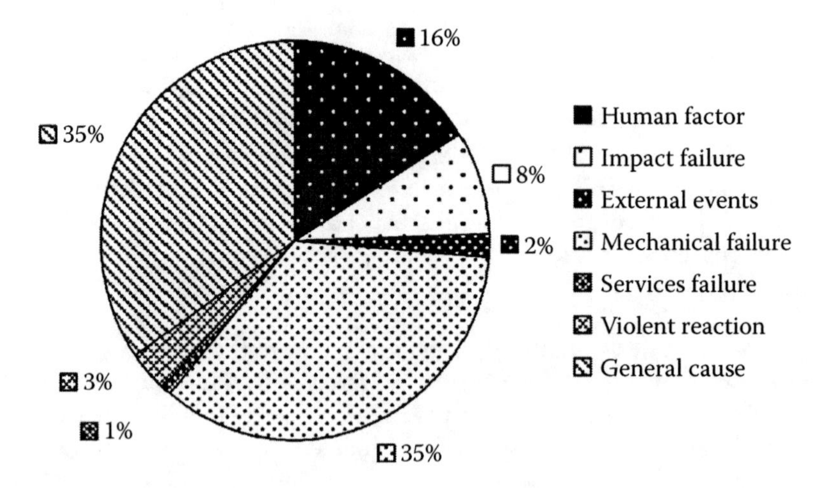

FIGURE 10.3
Causation factors for hydrogen accidents drawn from analysis of Major Hazard Incident Database Service (MHIDAS). (From Gerboni, R., and Salvador, E., *Energy*, 34 (12), 2223–2229, 2009. With permission.)

10.4 Other Examples

To conclude this chapter on case studies, examples are now given of incidents involving hydrogen in conjunction with other hazardous materials. We first examine the issue of hydrogen as a flammable gas coexisting with an explosible dust; this is followed by a brief look at the hazards associated with hydrogen being chemically bound with sulfur in the form of hydrogen sulfide.

10.4.1 Hydrogen and Explosible Dusts

A dust explosion can occur when particulate solid material is suspended in air and a sufficiently energetic ignition source is present within a confined or semiconfined environment [33]. In other words, dust explosions occur when the components of the explosion pentagon are all present: fuel, oxidant, ignition source, mixing, and confinement. The following two paragraphs, adapted from Amyotte and Eckhoff [33], provide additional background for the discussion here on explosions of hydrogen and combustible dusts.

Dust explosions usually occur in industry inside process vessels and units such as mills, grinders, and driers, that is, inside equipment where the conditions of the explosion pentagon are satisfied. Such occurrences are often called *primary* explosions, especially if they result in *secondary* explosions external to the process unit (as described in the next paragraph). The reason for the majority of dust explosions being initiated in this manner is relatively straightforward; the range of explosible dust concentrations in air at normal temperature and pressure is orders of magnitude greater than airborne dust concentrations permitted in areas inhabited by workers (i.e., in the context of industrial hygiene).

Notwithstanding the discussion in the previous paragraph, dust explosions do occur in process areas, not just inside process units. A *secondary* explosion can be initiated due to entrainment of dust layers by the blast waves arising from a *primary* explosion. The primary event might be a dust explosion originating in a process unit, or could be any disturbance energetic enough to disperse explosible dust layered on the floor and various work services. An example of such an energetic disturbance (other than a primary dust explosion) would be a gas explosion leading to a dust explosion. This is a well-documented phenomenon in the coal mining industry, where devastating effects can result from the overpressures and rates of pressure rise generated in a coal dust explosion that has been triggered by a methane explosion. (See Section 10.3 and the description of the Westray coal mine explosion.)

The above scenario appears to have been realized recently at the Hoeganaes facility in Gallatin, Tennessee. On May 27, 2011, an explosion and ensuing fire resulted in the death of two workers and significant injuries to a third [40]. Two previous incidents involving flash fires (solely iron dust) had occurred at this facility during 2011, resulting in the death of two workers on January 31 and injuries to another on March 29 [40]. As described in the CSB news conference statement [40] prepared following the May 27 incident, the Gallatin Hoeganaes facility manufactures atomized iron powder for the production of metal parts in the automotive and other industries. Hydrogen is used in the continuous annealing furnaces to prevent oxidation of the iron powder.

The initial explosion in the May 27 incident was caused by the ignition of hydrogen leaking into a process-pipe trench from a 7.6-cm by 17.8-cm hole in a vent pipe [40] (see Figure 10.4). This primary explosion disturbed layered iron dust (Figure 10.5), which subsequently burned in suspension as a flash fire—the secondary event.

At the time of writing, the CSB was continuing its investigation into all three incidents at the Hoeganaes facility with the aim of producing definitive conclusions as to root causes at the management-system level. Surely these conclusions will comment on the need for effective housekeeping programs to minimize dust layers and deposits. Lessons will also be learned from the May 27 incident with respect to the following hydrogen safety issues: (1) hydrogen gas alarms, (2) automatic shut-off systems, (3) maintenance and inspection of hydrogen piping, (4) procedures and training for response to flammable gas leaks, and (5) the adequacy of national codes for pipe maintenance and leak detection [40].

The discussion to this point has focused on series-type events, that is, a hydrogen explosion followed by a dust explosion (or fire). What happens when these events occur in parallel, that is, when hydrogen and dust ignition occur simultaneously? This scenario of admixture of a flammable gas

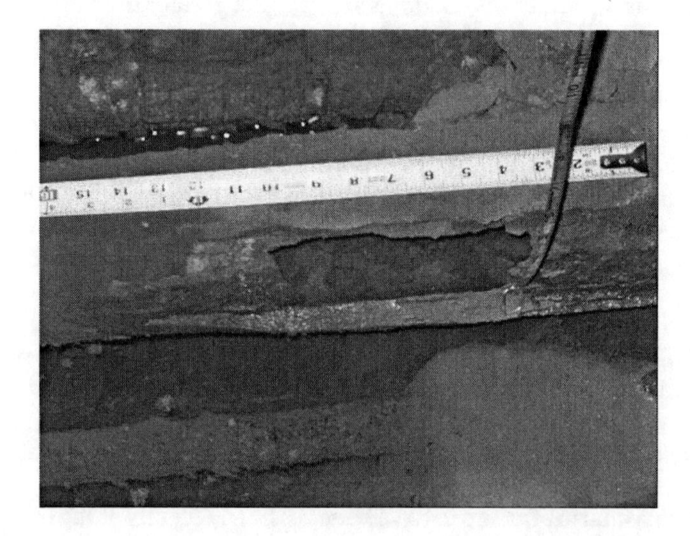

FIGURE 10.4
Photograph of a hole in hydrogen piping at Hoeganaes. (From the U.S. Chemical Safety Board web site, http://www.csb.gov/assets/news/image/Photo_of_hole_in_hydrogen_piping.JPG.)

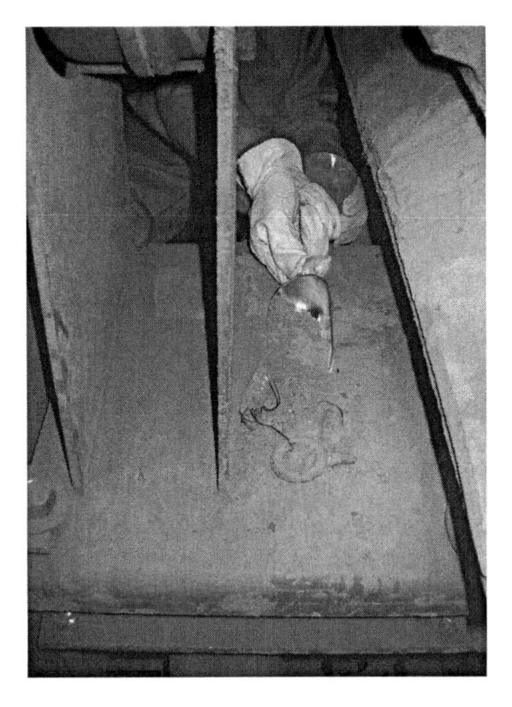

FIGURE 10.5
Photograph of layered iron dust deposits at Hoeganaes. (From the U.S. Chemical Safety Board web site, http://www.csb.gov/assets/Investigation/original/Dust_photo.jpg.)

and an atmosphere sustaining combustion of a dust is known as the creation of a hybrid mixture. Each component of the mixture (flammable gas and combustible dust) may be present in an amount less than its lower limit of reactivity (LFL or lower flammable limit for the gas, and MEC or minimum explosible concentration for the dust), and still give rise to an explosible mixture [33].

With the unfolding of the Fukushima disaster, the world was made aware of the potential for hydrogen explosions in nuclear facilities when exposed zircaloy fuel cladding comes in contact with steam. In a somewhat similar vein, a series of recent papers [43–45] has addressed the issue of combined hydrogen and graphite/tungsten dust explosions in a next-generation fusion machine known as the International Thermonuclear Experimental Reactor (ITER). As described by Chuyanov and Topilski [43], interaction of reactor components over time can be expected to liberate beryllium, graphite, and tungsten dust having characteristic diameters less than 100 μm. In the event of a water leak leading to contact between water and hot beryllium particles and the subsequent generation of hydrogen, the possibility exists for hydrogen/graphite/tungsten explosions. Thus, although this situation has not occurred, such an incident could further intensify the spotlight on hydrogen safety.

FIGURE 10.6
TEM (Tunneling Electron Microscope) photograph of 4-μm graphite dust (from Denkevits and Dorofeev [44]).

With respect to ITER-related dusts, Figure 10.6 illustrates the lower boundary of graphite particle size (4 μm) examined by Denkevits and Dorofeev [44] in their research on graphite/tungsten explosibility. This work enabled a baseline dataset to be established for graphite dust alone (4–45 μm), tungsten dust alone (1 μm), and graphite (4 μm)/tungsten (1 μm) mixtures. A follow-up study by Denkevits [45] investigated the explosibility of hybrid mixtures of the 4-μm graphite and lean hydrogen-air mixtures having hydrogen concentrations of 8 to 18 percent by volume. While valuable technical information was gained on the explosion behavior of the various fuel/air systems studied, the broadly applicable lesson to be learned from this case is the importance of credible scenario selection during the performance of risk assessments. Inadequate explosion prevention and mitigation measures could well be the result of neglecting a significant component of the total fuel loading.

10.4.2 Hydrogen Sulfide

Hydrogen sulfide (H_2S) is a colorless, toxic, and flammable gas with the characteristic odor of rotten eggs. Hydrogen sulfide is therefore a hazardous material; but it is not hydrogen. Thus, it may seem odd to include a discussion of hydrogen sulfide in a book on hydrogen safety. Our intention, however, is not to give a treatise on the hazards of hydrogen sulfide. Rather, we endeavor to illustrate, by means of examples involving H_2S, the importance of recognizing perception as an often-forgotten component of risk—in addition to the traditional components of likelihood and consequence severity. Word association alone may be sufficient for some people to perceive a wider risk than is actually present. Thus, it can be argued that safety practitioners concerned with hydrogen should be somewhat familiar with the lessons from selected case studies involving hydrogen sulfide.

The first example presented here comes from the (hopefully) familiar suite of investigations conducted by the U.S. Chemical Safety Board. Reference [46] reports the results of the CSB investigation into the 2002 hydrogen sulfide release at the Georgia-Pacific paper mill in Pennington, Alabama. H_2S gas generated in a sewer leaked from a manway seal, resulting in two deaths from poisoning and several injuries to contract workers and emergency responders. Several root and contributing causes were identified (all clearly at the management level), including (1) the lack of a formal hazard review and management of change analysis, (2) lack of a management system to incorporate hazard warnings into process safety information, (3) inadequate manway design and sealing, and (4) inadequate training to facilitate understanding hydrogen sulfide hazards [46].

The second example comes from a quite different industrial sector. Reference [47] reports the results of legal proceedings involving occupational health and safety charges related to an incident at a mushroom farm in Langley, British Columbia, in Canada. In September 2008, workers were attempting to repair an electric motor that powered a pump system located in a confined space, when a pump flange separated from the system resulting in a release of hydrogen sulfide [47]. Three people died from poisoning and two more were seriously injured [47]. Although this case again illustrates the need for an organized safety program [47], the key lesson to be learned from this incident is the need to employ the full hierarchy of safety controls (Section 7.1) when conducting work in confined spaces—starting with the inherently safer measure of eliminating such work whenever possible.

The final example presented in this chapter involves a crane incident at an oil processing facility. Figure 10.7 shows a crane that had toppled over because the boom had been left extended and unattended. In addition to damage caused by the boom-end of the crane (not shown in Figure 10.7), the cab-end resting place constitutes a near-miss with respect to the sour gas and refinery gas pipelines on the exposed pipe rack. There was significant potential in this incident for either a fire or hydrogen sulfide release. The importance of thorough hazard assessment is again emphasized by this case; in addition to dropped loads and crane vehicular accidents, it is imperative that the potential energy stored in an extended crane boom be carefully evaluated. In essence, this is an example of the need to employ the inherently safer design principle of moderation (Section 7.4).

FIGURE 10.7
Photograph of toppled crane at oil processing facility (from [48]).

References

1. Crowl, D.A., and Louvar, J.F., *Chemical Process Safety: Fundamentals with Applications*, 3rd edition, Prentice Hall PTR, Upper Saddle River, NJ, 2011.
2. Kletz, T., and Amyotte, P., *Process Plants: A Handbook for Inherently Safer Design*, 2nd edition, CRC Press/Taylor & Francis Group, Boca Raton, FL, 2010.
3. Stuart, C., Hydrogen leak in Connecticut forces dozens from homes, *New York Times* (February 13, 2008).
4. Leary, W.E., Hopes rise on fixing shuttle leaks so the fleet can resume flying, *New York Times* (July 14, 1990).
5. Broad, W.J., Shuttle astronomy mission is postponed by fuel leak, *New York Times* (September 6, 1990).
6. Leary, W.E., Fuel leak delays launching of space shuttle, *New York Times* (April 5, 2002).
7. Hu, W., Leak at Indian Point 2 plant leads it to curtail operations, *New York Times* (September 12, 2002).
8. International Conference on Hydrogen Safety (ICHS), http://www.ichs2011.com/themestopics. htm (accessed August 17, 2011).
9. Sanders, R.E., Designs that lacked inherent safety: Case histories, *Journal of Hazardous Materials*, 104 (1–3), 149–161, 2003.

10. Weiner, S.C., Fassbender, L.L., and Quick, K.A., Using hydrogen safety best practices and learning from safety events, *International Journal of Hydrogen Energy*, 36 (4), 2729–2735, 2011.
11. Kletz, T., *What Went Wrong? Case Histories of Process Plant Disasters and How They Could Have Been Avoided*, 5th edition, Gulf Professional Publishing, Oxford, UK, 2009.
12. Gupta, R.B. (Editor), *Hydrogen Fuel: Production, Transport, and Storage*, CRC Press/Taylor & Francis Group, Boca Raton, FL, 2009.
13. Hopkins, A., *Safety, Culture and Risk: The Organizational Causes of Disasters*, CCH Australia Limited, Sydney, Australia, 2005.
14. Hopkins, A. (Editor), *Learning from High Reliability Organisations*, CCH Australia Limited, Sydney, Australia, 2009.
15. *Hydrogen Safety for Employees*, http://www.safetyinstruction.com/hydrogen_safety_for_ employees_power_point.htm (accessed August 18, 2011).
16. *Brown McFarlane Case Studies*, http://www.brownmac.com/media-downloads/case-studies /?ID=18 (accessed August 18, 2011).
17. U.S. Department of Energy Hydrogen Program, http://www.hydrogen.energy.gov/permitting /fueling_case_studies_washington.cfm (accessed August 18, 2011).
18. Sanderson, K., Explosion at ammonia plant, *Chemistry World* (June 1, 2006).
19. Evans, J., Weapons of mass degradation, *Chemistry World* (2011).
20. Ramsey, B., Big batteries? Big electric shock potential! *Occupational Health & Safety*, (February 1, 2004).
21. Carson, P., and Mumford, C., Batch reactor hazards and their control, *Loss Prevention Bulletin*, 171, 13–24, 2003.
22. Institute of Chemical Engineers (IChemE), http://www.icheme.org/sitecore/content/icheme_home/shop/search results.aspx? keywords=hydrogen&product=&CurrentPage=1&SortBy=Relevance&OrderBy=Asce (accessed August 18, 2011).
23. U.S. Chemical Safety Board (CSB), http://www.csb.gov/ (accessed August 18, 2011).
24. U.S. Chemical Safety Board (CSB), *Positive Material Verification: Prevent Errors During Alloy Steel Systems Maintenance*, Safety Bulletin, No. 2005-04-B, U.S. Chemical Safety and Hazard Investigation Board, Washington, D.C., 2006.
25. Litterick, D., Texas refinery fire inquiry critical of BP, *The Telegraph* (October 17, 2006).
26. Mahnken, G.E., Use case histories to energize your HAZOP, *Chemical Engineering Progress*, 73–78, 2001.
27. Khan, F.I., and Amyotte, P.R., How to make inherent safety practice a reality, *Canadian Journal of Chemical Engineering*, 81 (1), 2–16, 2003.
28. Etowa, C.B., Amyotte, P.R., Pegg, M.J., and Khan, F.I., Quantification of inherent safety aspects of the Dow indices, *Journal of Loss Prevention in the Process Industries*, 15 (6), 477–487, 2002.
29. West, A.S., Hendershot, D., Murphy, J.F., and Willey, R., Bhopal's impact on the chemical industry, *Process Safety Progress*, 23 (4), 229–230, 2004.
30. Willey, R., Hendershot, D., and Berger, S., The accident in Bhopal: Observations 20 years later, *Process Safety Progress*, 26 (3), 180–184, 2007.

31. Louvar, J.F., Editorial. Bhopal, CCPS and 25 years, *Process Safety Progress*, 28 (4), 299, 2009.
32. *Bhopal Memorial*, http://upload.wikimedia.org/wikipedia/commons/d/d8/Bhopal-Union_ Carbide_1.jpg (accessed September 8, 2011).
33. Amyotte, P.R., and Eckhoff, R.K., Dust explosion causation, prevention and mitigation: An overview, *Journal of Chemical Health and Safety*, 17 (1), 15–28, 2010.
34. Richard, K.P., The Westray story: A predictable path to disaster, Report of the Westray Mine Public Inquiry, Province of Nova Scotia, Canada, 1997.
35. Sepeda, A.L., Lessons learned from process incident databases and the Process Safety Incident Database (PSID) approach sponsored by the Center for Chemical Process Safety, *Journal of Hazardous Materials*, 130 (1–2), 9–14, 2006.
36. Gerboni, R., and Salvador, E., Hydrogen transportation systems: Elements of risk analysis, *Energy*, 34 (12), 2223–2229, 2009.
37. Imamura, T., Mogi, T., and Wada, Y., Control of the ignition possibility of hydrogen by electrostatic discharge at a ventilation duct outlet, *International Journal of Hydrogen Energy*, 34 (6), 2815–2823, 2009.
38. Astbury, G.R., and Hawksworth, S.J., Spontaneous ignition of hydrogen leaks: A review of postulated mechanisms, *International Journal of Hydrogen Energy*, 32 (13), 2178–2185, 2007.
39. Kirchsteiger, C., Vetere Arellano, A.L., and Funnemark, E., Towards establishing an International Hydrogen Incidents and Accidents Database (HIAD), *Journal of Loss Prevention in the Process Industries*, 20 (1), 98–107, 2007.
40. U.S. Chemical Safety Board (CSB), http://www.csb.gov/assets/news/document/Final_Statement_6_3_2011.pdf (accessed September 8, 2011).
41. U.S. Chemical Safety Board (CSB), http://www.csb.gov/assets/news/image/Photo_of_hole_in_hydrogen_piping.JPG (accessed September 8, 2011).
42. U.S. Chemical Safety Board (CSB), http://www.csb.gov/assets/Investigation/original/Dust_photo.jpg (accessed September 8, 2011).
43. Chuyanov, V., and Topilski, L., Prevention of hydrogen and dust explosion in ITER, *Fusion Engineering and Design*, 81 (8–14), 1313–1319, 2006.
44. Denkevits, A., and Dorofeev, S., Explosibility of fine graphite and tungsten dusts and their mixtures, *Journal of Loss Prevention in the Process Industries*, 19 (2–3), 174–180, 2006.
45. Denkevits, A., Explosibility of hydrogen-graphite dust hybrid mixtures, *Journal of Loss Prevention in the Process Industries*, 20 (4–6), 698–707, 2007.
46. U.S. Chemical Safety Board (CSB), *Hydrogen Sulfide Poisoning*, Investigation Report, No. 2002-01-I-AL, U.S. Chemical Safety and Hazard Investigation Board, Washington, D.C., 2003.
47. Contant, J., Guilty pleas in confined space deaths, *OHS Canada*, 27 (5), 6, 2011.
48. *Crane Incident*, http://farm1.static.flickr.com/34/103084821_c1066589e1_b.jpg (accessed September 8, 2011).

11

Effects of Hydrogen on Materials of Construction

Chapter 4 presented in detail the primary physiological, physical, and chemical hazards of hydrogen. In this chapter we return briefly to the issue of the physical hazards of hydrogen from the perspective of the subsequent effects on materials of construction. To recapitulate, Chapter 4 categorized these physical hazards as being related to either hydrogen's small molecular size or its storage at generally low temperatures. Two key issues arise from these factors: degradation of structural materials because of hydrogen embrittlement, and loss of thermal stability, respectively.

From a functional point of view, broad advice on the effects of hydrogen on materials of construction is given in a number of references. A self-study training guide on hydrogen gas safety prepared by Los Alamos National Laboratory [1] reiterates the above point that due to its small molecular size, hydrogen is able to readily pass through porous materials and be absorbed by some containment materials (the end result being loss of ductility, or embrittlement, at a rate that is temperature dependent). The guide therefore recommends the selection of materials for use with hydrogen that are not subject to degradation by hydrogen embrittlement (such as 300-series stainless steels, copper, and brass) [1]. Similar guidance is given by Molkov [2], who cautions on the decarburization (decrease in carbon content) and embrittlement of mild steels caused by exposure to hydrogen at elevated temperature and pressure. Selection of appropriate materials such as special alloy steels is critical to address the issue of hydrogen embrittlement [2].

Metals are not the only structural materials for which hydrogen-related issues may arise. As noted in the HySafe work package on material compatibility and structural integrity (see Section 9.2 and reference [3]), hydrogen components and hydrogen systems also involve nonmetallic materials such as polymers. HySafe researchers have recommended that materials intended for hydrogen service be carefully evaluated for use under the design, operating, and emergency conditions to which they will be exposed [3]. This includes storage conditions such as solid hydrides based on nano-structured carbonaceous materials [3] for which hazard uncertainties exist [3, 4].

All of the above points in this introductory section are consistent with the main materials-related findings given in a key report produced by the Health and Safety Laboratory in the United Kingdom [5]:

- Primary material considerations with hydrogen usage are permeability, hydrogen embrittlement, and properties of materials at low (cryogenic) temperatures typically involved in the storage and use of liquid hydrogen.
- Increased hydrogen prominence will likely result in the use of new materials, for example, high-performance composites for pressurized hydrogen cylinders, combinations of materials for pipelines, and materials such as metal hydrides and carbon nanotubes for hydrogen storage.
- Materials proposed for use with hydrogen must be tested in the environment in which they will actually be employed. (Reference [5] notes that this requirement may necessitate the use of specialist equipment and techniques.)

This report by Hobbs [5] is also of interest because it extends the discussion on material-related effects to considerations beyond the physical hazards of hydrogen. By drawing on a NASA (U.S. National Aeronautics and Space Administration) document [6] (also reference [6] in Chapter 4), Hobbs [5] lists 12 aspects to be considered when selecting materials for use with hydrogen. The following excerpted points are as relevant in 2011 as they were in 2005 [5] and also in 1997 [6]; account should be taken of:

- Properties suitable for the design and operation conditions
- Compatibility with the operating environment
- Availability of selected materials and appropriate test data
- Corrosion resistance
- Ease of fabrication, assembly, and inspection
- Consequences of failure
- Toxicity
- Hydrogen embrittlement
- Potential for exposure to high temperature from a hydrogen fire
- Cold embrittlement
- Thermal contraction
- Property changes that occur at cryogenic temperatures

In the above list we see not only the familiar topics of hydrogen embrittlement and cryogenic-temperature property changes, but also considerations related to material availability and failure consequence analysis. These are important issues with respect to inherently safer design substitutions (Section 7.3) and process risk management (Section 8.2.4), respectively. From a general process safety management perspective, and again in accordance with the discussion in Chapter 4, Hobbs [5] comments that these points apply to all hydrogen-service components, not just the more obvious ones such as containment vessels. Such components include valve bodies and seats, electrical systems, insulation, gaskets, seals, tubing, lubricants, and adhesives [5].

11.1 Hydrogen Embrittlement

Chapter 4 covered in some detail various elements of hydrogen embrittlement, including types, causes, and rate-influencing parameters. Here we present information of an applied, example-based nature aimed at a general understanding of hydrogen embrittlement impacts on constructional materials.

The article by Still [7] is helpful in this regard, with a particular emphasis on the offshore oil and gas industry; the author uses the example of an FPSO (floating, production, storage, and offloading) vessel consisting of piping systems connecting subsea reservoirs to onboard processing equipment. Two main sources of hydrogen ingress are described as being problematic for carbon and carbon-manganese steel pipelines and weld metals: (1) atomic hydrogen ingress resulting from the processing of reservoir fluids, and (2) because of the presence of moisture, hydrogen introduction into the weld pool during welding of process piping [7].

Still [7] describes in detail the various types of hydrogen embrittlement shown in Table 4.1, including *environmental hydrogen embrittlement* (termed HSC or hydrogen stress cracking). Also elucidated is a form of *internal hydrogen embrittlement* (Table 4.1) known as a *fisheye* (Figure 11.1). Fisheyes are described as a type of hydrogen embrittlement occurring in as-welded structures, wherein hydrogen is retained within a weld defect such as a pore [7]. Still [7] emphasizes the need to adhere to standardized welding controls to avoid this form of hydrogen failure.

An example of *hydrogen reaction embrittlement* (Table 4.1) is given in the CSB safety bulletin [8] describing an incident at the BP Texas City refinery. This incident was previously discussed in Sections 7.6, 8.2, and 10.1 from the perspectives of inherently safer design principles, safety management systems, and case studies, respectively. Here the focus is on the material damage caused by high-temperature hydrogen attack or HTHA.

The following excerpt from reference [8] provides incident background and consequence details:

> On July 28, 2005, 4 months after a devastating incident in the Isomerization (Isom) unit that killed 15 workers and injured 180, the BP Texas City refinery experienced a major fire in the Resid Hydrotreater Unit (RHU) that caused a reported $30 million in property damage. One employee sustained a minor injury during the emergency unit shutdown and there were no fatalities.

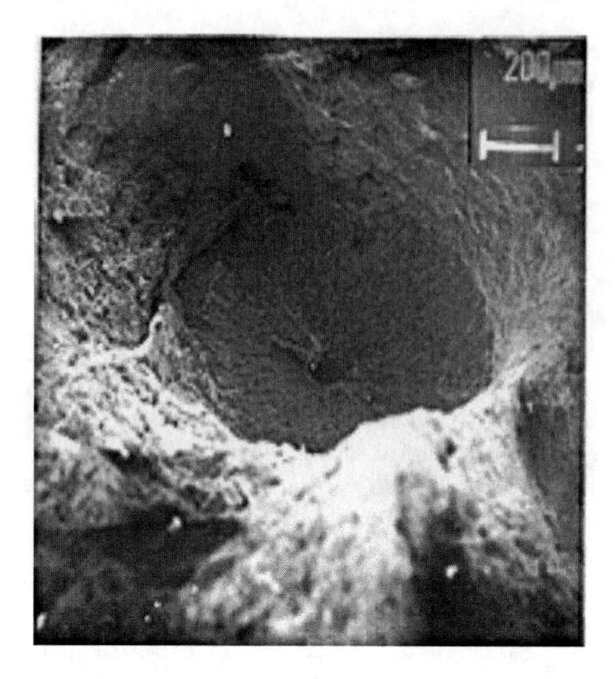

FIGURE 11.1
Example of a fisheye, a form of hydrogen embrittlement. (From Still, J.R., *Welding Journal*, January, 2004. With permission. http://www.aws.org.wj/jan04/still_feature.html.)

> The RHU incident investigation determined that an 8-inch diameter carbon steel elbow inadvertently installed in a high-pressure, high-temperature hydrogen line ruptured after operating for only 3 months. The escaping hydrogen gas from the ruptured elbow quickly ignited.
>
> This incident occurred after a maintenance contractor accidentally switched a carbon steel elbow with an alloy steel elbow during a scheduled heat exchanger overhaul in February 2005. The alloy steel elbow was resistant to high temperature hydrogen attack (HTHA) but the carbon steel elbow was not. Metallurgical analyses of the failed elbow concluded that HTHA severely weakened the carbon elbow steel.

Figure 11.2 illustrates the potential that existed at this facility for human error to occur with respect to elbow replacement. As discussed in Section 8.2.7, this potential is most effectively countered by human factor considerations during process design—in this case by a design that avoids configurations that allow critical alloy piping components to be interchanged with incompatible piping components [8].

The CSB report [8] remarks that the occurrence of incidents involving HTHA has been documented since the 1940s. It is known that carbon steel is vulnerable to HTHA at pressures above 100 psia (7 bar(a)) and temperatures above about 450°F (230°C); these conditions promote the permeation of hydrogen through the steel and the reaction of hydrogen with carbon and carbides to produce methane gas [8]. This loss of carbon is known as decarburization (as earlier described [2]) and results in steel degradation and eventually, in the case of this particular incident, pipeline rupture [8]. These effects of hydrogen reaction embrittlement (in particular, high-temperature hydrogen attack) are shown in Figure 11.3.

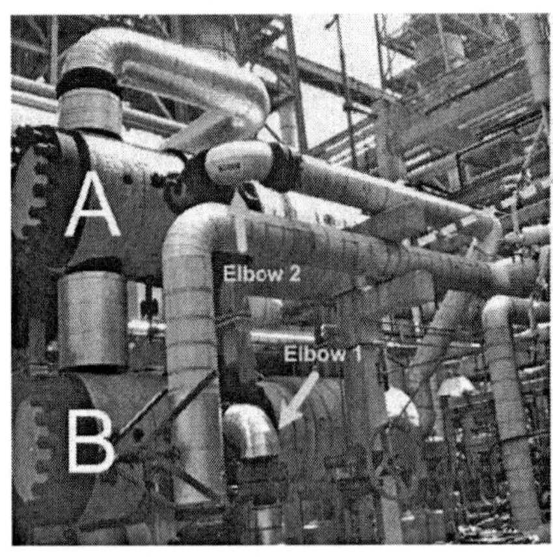

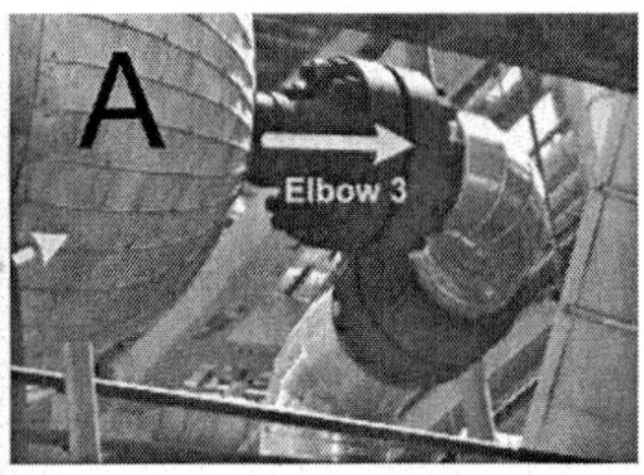

Upper Left & Top Arrow – Alloy steel elbows 2 and 3
Lower Left Arrow – Carbon steel elbow 1

FIGURE 11.2
Heat-exchanger piping elbows with identical dimensions. (From U.S. Chemical Safety Board (CSB), *Positive Material Verification: Prevent Errors During Alloy Steel Systems Maintenance*, Safety Bulletin, No. 2005-04-B, U.S. Chemical Safety and Hazard Investigation Board, Washington, D.C., 2006. With permission.)

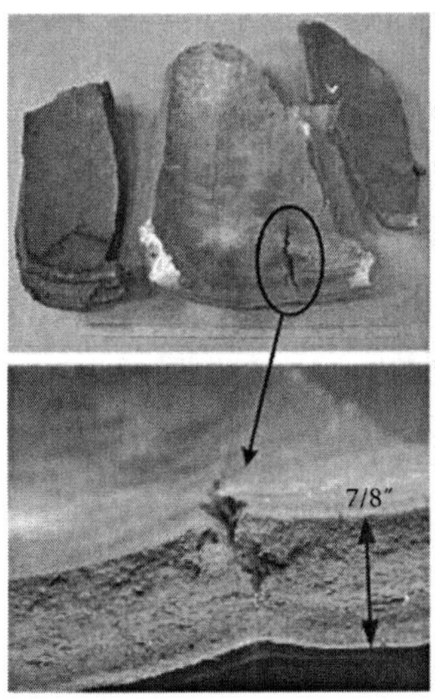

Upper Left – Carbon steel elbow segments (view of inside surface)
Above – Flange segments
Left – Close-up of fissure on middle elbow segment

FIGURE 11.3
Ruptured carbon-steel pipe elbow subjected to high-temperature hydrogen attack (from CSB [8]).

11.2 Loss of Thermal Stability

The discussion in Chapter 4 on loss of thermal stability focused on low-temperature embrittlement and thermal contraction at cryogenic temperatures. Hobbs [5] reiterates the familiar concept that because of the low temperatures at which liquid hydrogen is stored, special attention must be paid to the properties of the materials used for such storage. A primary concern is the possibility of transition from ductility to brittleness as temperature decreases. This was addressed quantitatively in Chapter 4 by the Charpy impact data shown in Figure 4.1; these data clearly demonstrate the preferential use of 304 stainless steel over carbon steel and 201 stainless steel for low-temperature hydrogen service.

The importance of proper material selection for operations involving low temperatures is amply demonstrated by the Longford gas plant explosion in Australia. Although not a liquid hydrogen storage incident, as an additional case study of lessons to be learned from process incidents the treatment of Longford by sociologist Andrew Hopkins [9] is a key resource for understanding the potential consequences of a loss of thermal stability (from both perspectives of low-temperature embrittlement and thermal contraction).

It should be noted that the focus of Hopkins's analysis is on incident causation due to organizational failures such as the safety management system and safety culture deficiencies [9]. Readers are directed to reference [9] for a full exposition of this analysis and the resulting generalized lessons. For our purposes here, the following excerpt provides sufficient detail on loss of thermal stability [9]:

> Because the circulation of warm lean oil had stopped, two of the heat exchangers became abnormally cold and a thick layer of frost formed on their exterior pipework. The temperature dropped below the design limit and *the metal in one exchanger contracted to the point that it began to leak oil onto the ground*. Unsuccessful attempts were made to fix this leak by tightening certain bolts. Operators decided to stop the flow into GP 1 [Gas Plant 1] at this point to try to deal with the situation. This stopped any further flow of cold condensate within the plant. But operators did not depressurize the plant. Rather they decided to try again to restart the pumps to rewarm the heat exchanger. This was a critical error. *The metal in the vessel was by this time so cold that it was brittle* and it needed time to thaw out before being rewarmed. Operators succeeded in restarting the pumps and the reintroduction of warm liquid caused fracturing and catastrophic failure of one of the heat exchangers. A large quantity of volatile liquid and gas escaped and was ignited by a nearby ignition source.

The passages emphasized with italics (author added) in the above excerpt give practical examples of, respectively, thermal contraction and low-temperature embrittlement (cold-temperature or cold-metal embrittlement as it is termed in reference [9]). The ultimate consequences of these physical phenomena occurring at Longford include two fatalities, extensive asset damage, and significant business interruption for the plant owner and the surrounding communities.

11.3 Ongoing Research

While practical information such as that derived from case studies is important in reducing hydrogen storage risks, fundamental research is essential to elucidate corrosion mechanisms and embrittlement reaction pathways. Obtaining a better understanding of such features is the first step in designing enhanced prevention and mitigation techniques.

To conclude, Table 11.1 was developed from a search of the archival publications *Corrosion Science* and the *International Journal of Hydrogen Energy*. Table 11.1 is not intended as a comprehensive listing, but rather as a snapshot

TABLE 11.1

Examples from the Literature of Research on the Effects of Hydrogen on Material Properties

Reference	Research Focus
Smiyan et al. [10]	Causes and mechanism of initiation and growth of corrosion sites on the internal surface of the shell of high-pressure steam boilers
Torres-Islas et al. [11]	Effects of hydrogen on the mechanical properties of X70 pipeline steel at different heat treatments
Rogante et al. [12]	Intercrystalline and intergranular crack detection in narrow welded zones of pipelines
Nikiforov et al. [13]	Corrosion resistance of commercially available stainless steels, Ni-based alloys, titanium, and tantalum under conditions corresponding to those in high-temperature proton exchange membrane (PEM) steam electrolyzers
Figueroa and Robinson [14]	Feasibility of zinc/nickel and aluminum-based coatings as replacements for cadmium in steel-coating processes
Michler et al. [15]	Measurement of environmental hydrogen embrittlement of several heats of austentitic stainless steels by slow strain rate tensile testing at various combinations of temperature and pressure
Kittel et al. [16]	Investigation of hydrogen-induced cracking of pipeline steel by means of immersion testing and hydrogen permeation measurements
Capelle et al. [17]	Assessment of hydrogen absorption abilities of X52, X70, and X100 pipeline steels through consideration of cathodic hydrogen charging, time of exposure, and applied stress
Yao et al. [18]	Effects of hydrogen on the fracture stress of passive films formed on 316L stainless steel
Venegas et al. [19]	Resistance to hydrogen-induced cracking of low-carbon pipeline steel samples by control of crystallographic texture and grain-boundary distribution

of some of the recent and ongoing research related to properties of materials for use with hydrogen. This is clearly an important area of hydrogen safety that will benefit from continued effort as new opportunities for hydrogen usage arise and new materials for hydrogen storage emerge.

References

1. Basquin, S., and Smith, K., *Hydrogen Gas Safety. Self-Study*, Document No. ESH13-401-sb-8/00, Los Alamos National Laboratory, Los Alamos, NM, 2000.
2. Molkov, V., Hydrogen safety research: State-of-the-art, in Proceedings of the 5th International Seminar on Fire and Explosion Hazards, Edinburgh, UK (April 23–27, 2007).
3. *HySafe – Safety of Hydrogen as an Energy Carrier*, http://www.hysafe.net/ (accessed September 22, 2011).
4. Amyotte, P.R., *Are Classical Process Safety Concepts Relevant to Nanotechnology Applications?* Journal of Physics: Conference Series 304 (Nanosafe2010: International Conference on Safe Production and Use of Nanomaterials), 012071, 2011.
5. Hobbs, J., *The Hydrogen Economy: Evaluation of the Materials Science and Engineering Issues*, HSL/2006/59, Health and Safety Laboratory, Buxton, UK, 2005.
6. NASA, *Safety Standard for Hydrogen and Hydrogen Systems*, National Aeronautics and Space Administration, Report NSS 1740.16, Office of Safety and Mission Assurance, Washington, D.C., 1997.
7. Still, J.R., Understanding hydrogen failures, *Welding Journal*, American Welding Society (January 2004), http://www.aws.org.wj/jan04/still_feature.html (accessed September 22, 2011).
8. U.S. Chemical Safety Board (CSB), *Positive Material Verification: Prevent Errors During Alloy Steel Systems Maintenance*, Safety Bulletin, No. 2005-04-B, U.S. Chemical Safety and Hazard Investigation Board, Washington, D.C., 2006.
9. Hopkins, A., *Lessons from Longford. The Esso Gas Plant Explosion*, CCH Australia Limited, Sydney, Australia, 2000.
10. Smiyan, O.D., Grigorenko, G.M., and Vainman, A.B., Effect of hydrogen on corrosion damage of metal of the high-pressure energetic boiler drum, *International Journal of Hydrogen Energy*, 27 (7–8), 801–812, 2002.
11. Torres-Islas, A., Salinas-Bravo, V.M., Albarran, J.L., and Gonzalez-Rodriguez, J.G., Effect of hydrogen on the mechanical properties of X-70 pipeline steel in diluted $NaHCO_3$ solutions at different heat treatments, *International Journal of Hydrogen Energy*, 30 (12), 1317–1322, 2005.
12. Rogante, M., Battistella, P., and Cesari, F., Hydrogen interaction and stress-corrosion in hydrocarbon storage vessel and pipeline weldings, *International Journal of Hydrogen Energy*, 31 (5), 597–601, 2006.
13. Nikiforov, A.V., Petrushina, I.M., Christensen, E., Tomas-Garcia, A.L., and Bjerrum, N.J., Corrosion behaviour of construction materials for high temperature steam electrolysers, *International Journal of Hydrogen Energy*, 36 (1), 111–119, 2011.

14. Figueroa, D., and Robinson, M.J., The effects of sacrificial coatings on hydrogen embrittlement and re-embrittlement of ultra high strength steels, *Corrosion Science*, 50 (4), 1066–1079, 2008.
15. Michler, T., Yukhimchuk, A.A., and Naumann, J., Hydrogen environment embrittlement testing at low temperatures and high pressures, *Corrosion Science*, 50 (12), 3519–3526, 2008.
16. Kittel, J., Smanio, V., Fregonese, M., Garnier, L., and Lefebvre, X., Hydrogen induced cracking (HIC) testing of low alloy steel in sour environment: Impact of time of exposure on the extent of damage, *Corrosion Science*, 52 (4), 1386–1392, 2010.
17. Capelle, J., Dmytrakh, I., and Pluvinage, G., Comparative assessment of electrochemical hydrogen absorption by pipeline steels with different strength, *Corrosion Science*, 52 (5), 1554–1559, 2010.
18. Yao, Y., Qiao, L.J., and Volinsky, A.A., Hydrogen effects on stainless steel passive film fracture studied by nanoindentation, *Corrosion Science*, 53 (9), 2679–2683, 2011.
19. Venegas, V., Caleyo, F., Baudin, T., Espina-Hernandez, J.H., and Hallen, J.M., On the role of crystallographic texture in mitigating hydrogen-induced cracking in pipeline steels, *Corrosion Science*, In press, 2011. doi:10.1016/j.corsci.2011.08.031.

12

Future Requirements for Hydrogen Safety

In this chapter we present some thoughts on areas requiring further research to more fully realize the anticipated benefits of the hydrogen economy. In this regard, Guy [1] notes that there are several issues concerning safe handling that must be successfully addressed before hydrogen usage gains wide public acceptance. These constraints have been described by Dahoe and Molkov [2] as involving both technical and nontechnical barriers. Nontechnical issues include the perception that hydrogen is a dangerous substance with a propensity for creating fire and explosion hazards; technical issues relate to the requirement to ensure hydrogen technologies have the same level of safety as those based on fossil fuels [2].

Salvi [3] extends the latter point above in his description of the European Union "Grand Challenges," one of which concerns sustainable energy and the greening of transport. According to Salvi [3], with italics added for emphasis:

> The safe development of clean energy alternatives [requires] that several problems have to be solved regarding the technology itself but also the integration of the new technology in the present systems; the greening of transport has to be treated with the same systemic view, e.g. the existing underground infrastructures have to be adapted to the new energy vehicles using *hydrogen*, batteries or compressed natural gas.

The following sections outline hydrogen research needs from several different perspectives: public safety, occupational safety, and process safety. Section 12.1 is as brief review of the literature, followed, in Section 12.2, by a discussion of research needs that we have identified.

12.1 Hydrogen Safety Research Gaps Identified in the Literature

In their commentary on hydrogen knowledge gaps, Pasman and Rogers [4] consider two primary sources of information: (1) the Hydrogen Research Advisory Council working under the umbrella of the Fire Protection Research Foundation (linked to the U.S. National Fire Protection Association, or

NFPA), and (2) the International Energy Agency (IEA) under the Hydrogen Implementing Agreement.

The NFPA-linked initiative identified 27 items, with the 11 topics below being the most urgent, in order of priority (as described by Pasman and Rogers [4]):

- Refinement of explosion models for flame speeds, blast waves, etc.
- Development and evaluation of wide-area sensing technology
- Effects on materials of construction (in particular, fatigue loading)
- Hydrogen gas cabinets (for storage of pressurized-gas cylinders)
- Deflagrations with partial confinement
- Reliability of pressure relief devices
- Strategies for mitigation of confined releases
- Design, installation, testing, and maintenance of hydrogen detection systems
- Ignition limits and criteria for large-leak, dynamic scenarios

Pasman and Rogers [4] further explain that much is unknown about hydrogen ignition probability and its dependence on release conditions and leak characteristics. This is a particularly important concern in risk assessment studies [4], specifically infrastructure safety studies and effectiveness of fire barriers.

The white paper produced under the IEA initiative categorized hydrogen safety requirements in three groups of knowledge gaps described by Pasman and Rogers [4] as related to:

- Existing codes and standards, as well as the ongoing development of such codes and standards
- Existing risk assessment methods and their application to hydrogen systems
- Fundamental knowledge (with a link to modeling approaches, including those employing computational fluid dynamics, or CFD)

Additional thoughts on unresolved items related to hydrogen safety can be found in the description of the achievements of the HySafe network of excellence (see Chapter 9) by Jordan et al. [5]. The following issues have been identified as requiring further research from a technical/scientific perspective [5]:

- Properties and behavior of cold, liquid hydrogen releases
- Mitigation strategies in response to accidental release scenarios (e.g., optimization of pressure-relief device placement and operation)
- Ignition phenomena (particularly in relation to modeling ignition probability)

- Impinging and wall-attached jets and jet fires (in relation to safe blowdown conditions)
- Sensor technologies
- Transitional combustion phenomena under realistic scenarios (e.g., low-temperature, congested environments with nonuniform mixing) and the associated impact on mitigation measures to deal with, for example, the processes of flame acceleration and deflagration-to-detonation transition (DDT) with water sprays
- Permitting requirements for hydrogen vehicle use in confined spaces
- Fundamental understanding of hydrogen behavior in confined spaces
- Reference quantitative risk assessment (QRA) methodology for application to scenarios involving garages and tunnels
- Reference simulation tool for combustion that is widely available to researchers
- Composite storage and vehicle safety testing strategies
- Pipeline field tests

One sees in the above list several items in common with the previous two from Pasman and Rogers [4], for example, ignition phenomena and probability modeling, risk assessment methodologies, and sensor/detection systems. Also evident in the list from Jordan et al. [5], as would be expected, is an emphasis on specific HySafe focus areas such as garages, tunnels, and confined spaces in general.

Two recent contributions from Vladimir Molkov of the University of Ulster, United Kingdom, provide further insight into hydrogen safety research needs from a technical perspective. The first is a review of the state of the art of hydrogen safety research [6], which identifies the following broad areas having several unresolved issues:

- Releases
- Ignitions
- Jet flames
- Premixed and partially premixed combustion
- Deflagration-to-detonation transition
- CFD for hazard and risk assessment

Reference [6] provides a detailed breakdown of research needs within each area, and there is considerable agreement between their list of research topics and that given by Jordan et al. [5]. One area of particular importance is the spontaneous ignition of pressurized hydrogen releases [6]. This is also evident in the nature of the submissions to a 2008 special issue on hydrogen safety of the *Journal of Loss Prevention in the Process Industries*; in his preface to the issue, Molkov [7] comments that 4 of the 10 papers are related to the practical problem of spontaneous ignition of sudden high-pressure releases of hydrogen. The remaining papers deal with the familiar topics of hydrogen-air detonations, metal hydride fire incidents, CFD modeling, and hydrogen safety training and education [7]—all issues that are identified above.

The needs identified by other researchers and practitioners that have been summarized here can serve as an entry point to hydrogen safety research for those interested in the field. Although other aspects of hydrogen research dealing with technical feasibility issues are of obvious importance, the general theme of this book has been to illustrate the overarching need for complementary work on safety of hydrogen as an energy carrier.

12.2 Hydrogen Safety Needs: Some Thoughts on Research

The concluding chapter of this book presents a brief recapitulation of our thoughts on the need for continued research on hydrogen safety. The process industries have a decades-old track record of handling hydrogen in a manner that attempts to reduce the risk to an acceptable level. While process incidents involving hydrogen have indeed occurred (see Chapter 10), it would seem helpful to look to this industry sector for guidance on future research needs.

Reference [8], on process safety research needs from an industrial perspective, was written to stimulate discussion at a workshop "Frontiers of Research," organized by Professor Sam Mannan of the Mary Kay O'Connor Process Safety Center at Texas A&M University. The remainder of this chapter is adapted from Amyotte [8], taking hydrogen-specific issues into consideration.

12.2.1 General Process Safety Research Needs

Assuming that one's own process safety research topics represent true industrial needs may seem somewhat self-centered. Perhaps this is a valid assumption, however, when the research is directly supported by industry and funded by granting agencies through strategic, targeted programs. With this reasoning, current research on dust explosions by Amyotte and Khan [9] can be viewed as industrially relevant.

This work is being conducted under four themes: (1) experimentation, (2) modeling, (3) development of risk management protocols, and (4) communication to industry, academia, and the general public. The fuel/air systems being studied experimentally include selected nanomaterials, flocculent (fibrous) materials, and hybrid mixtures (mixtures of a combustible dust and a flammable gas). The modeling work includes phenomenological, thermokinetic, and CFD approaches. The risk management component is an attempt to help advance the field of dust explosion prevention and mitigation from an emphasis on *hazards* (with an accompanying reliance on primarily engineered or add-on safety features) to a focus on *risk* (with an accompanying reliance on hierarchical, risk-based decision-making tools). The final theme is a recognition of the serious problem that exists with respect to effectively communicating the results of dust explosion research in a meaningful manner to various stakeholders.

Content-wise, what likely facilitated the funding of the above research are the following points:

- A choice of fuel/air systems for which there is industry consensus on the need for additional data.
- The coupling of experimental work with modeling research (so as to eventually lessen the very empirical nature of this field).
- A move away from absolute safety measures to relative measures selected on a risk basis.
- An expressed desire to communicate in ways other than journal papers and conference presentations (as important as these dissemination avenues are).

Again, it seems self-centered to promote the above points as universal factors for judging what constitutes research having industrial relevance. But it also seems difficult to argue that these points are pertinent only to the case presented here (Amyotte and Khan [9]) for illustrative purposes.

To broaden the search for industry's process safety research needs, a number of thematic papers on the topic (Amyotte [10], Hendershot et al. [11], Kletz [12], Knegtering and Pasman [13] and Qi et al. [14]) were consulted to identify recurrent keywords. Notice was also taken of the session titles from recent editions of the *Global Congress on Process Safety* sponsored by the Center for Chemical Process Safety (CCPS) and the Safety and Health Division of the American Institute of Chemical Engineers (AIChE). This event now consists of the *Loss Prevention Symposium, Process Plant Safety Symposium,* and *CCPS International Conference.* Also helpful in this regard were the *Hazards Symposia* of the Institution of Chemical Engineers and the *Loss Prevention and Safety Promotion in the Process Industries Symposia* of the European Federation of Chemical Engineering.

The above procedure resulted in the general keywords shown in the first column of Table 12.1. Papers incorporating these keywords were then identified by searching three archival journals whose scope is primarily related to process safety: (1) the *Journal of Loss Prevention in the Process Industries* (JLPPI, over the period 1988–2011), (2) *Process Safety and Environmental Protection* (PSEP, over the period 1996–2011), and (3) *Process Safety Progress* (PSP, over the period 1993–2011). The entries in the column for each journal in Table 12.1 give the number of papers for which the authors self-reported the given

TABLE 12.1

General Process Safety Keywords and Corresponding Journal Papers

Keywords	JLPPI	PSEP	PSP	IJHE
Risk assessment	110	39	21	13
Quantitative risk assessment	20	5	6	7
Safety management *or* Safety management system	41	10	29	1
Accident investigation *or* Incident investigation	7	4	7	0
Human error *or* Human factors	23	12	13	0
Safety culture	6	7	11	0
Performance indicator *or* Key performance indicator	3	2	7	2
Inherent safety *or* Inherently safer design	21	20	14	2
Security *or* Plant security	9	6	1	4
Offshore Safety	12	3	1	0
Reactive chemistry *or* Reactivity hazards	3	1	2	0
Dust explosion	113	10	9	1
Gas explosion	64	12	9	1

Note: JLPPI, *Journal of Loss Prevention in the Process Industries* (1988–2011); PSEP, *Process Safety and Environmental Protection* (1996–2011); PSP, *Process Safety Progress* (1993–2011); IJHE, *International Journal of Hydrogen Energy* (1976–2011), as discussed in Section 12.2.2.

keyword. Table 12.2 repeats this process for a set of more specific keywords that were identified in the thematic papers referenced in the previous paragraph. (Note that discussion of the final column in Tables 12.1 and 12.2, IJHE or *International Journal of Hydrogen Energy*, is deferred until Section 12.2.2.)

Comparisons among the data shown in Tables 12.1 and 12.2, as well as any conclusions drawn from the data, are fraught with potential pitfalls, such as those related to the following questions:

- Because a topic is the subject of extensive research, does that necessarily mean the topic is reflective of an actual industry need? While some of the papers appearing in JLPPI, PSEP, and PSP are written by industrial practitioners, many submissions to these journals come from researchers based in academia and government centers.
- What is the impact of special issues that are devoted to a particular topic? For example, JLPPI has published special issues consisting of papers presented at the *International Symposia on Hazards, Prevention, and Mitigation of Industrial Explosions* (ISHPMIE); these papers deal almost exclusively with gas and dust explosions. This undoubtedly helps to explain the high number of JLPPI papers listed in the last two rows of Table 12.1, and limits the usefulness of any attempts to normalize the data in Tables 12.1 and 12.2.
- Should one conclude that a topic showing low activity is not as applicable to industry as one having more papers that cite the topic as a keyword? As one example, is reactive chemistry/reactivity hazards

TABLE 12.2

Specific Process Safety Keywords and Corresponding Journal Papers

Keywords	JLPPI	PSEP	PSP	IJHE
HAZOP	18	13	8	1
FMEA	1	0	0	2
Fault tree	13	6	2	3
Event tree	5	4	0	1
Bow-tie	3	4	2	0
Bayesian	9[a]	3	0	1

Note: JLPPI, *Journal of Loss Prevention in the Process Industries* (1988–2011); PSEP, *Process Safety and Environmental Protection* (1996–2011); PSP, *Process Safety Progress* (1993–2011); IJHE, *International Journal of Hydrogen Energy* (1976–2011), as discussed in Section 12.2.2.

[a] All of these papers were published in 2006 or later, with the exception of Kirchsteiger [15].

research less important to industry than human error/human factors research? Such a conclusion drawn solely from the data in Table 12.1 would be speculative at best. At worst, such a conclusion could direct research effort in a manner that is counterproductive.

In spite of the difficulties just articulated, two key industry needs can be postulated and put forward for discussion based on Tables 12.1 and 12.2. First, continued research is required in the following five broad areas: (1) risk assessment (both qualitative and quantitative), (2) safety management systems (including various elements such as incident investigation and human factors), (3) safety culture (including the concept of key performance indicators or KPIs), (4) inherently safer design (including security as well as safety issues, and offshore as well as land-based facilities), and (5) material hazards (as determined by chemical reactivity, flammability, and explosibility). Second, both long-standing methodologies (e.g., fault trees) and newer techniques (e.g., Bayesian networks; see footnote to Table 12.2) have a role to play in advancing the body of knowledge within these five broad research areas.

The following additional points are offered concerning industry needs with respect to process safety research:

- Industry needs (or rather, *wants*) process safety analysis tools that are simple to use. The phrase *simple to use* should not be confused with *simplistic*. Scientific and engineering rigor are expected in all process safety research efforts. This may be an obvious statement, but it is one that bears repeating.
- In determining industry needs, we should consider case histories of process-related incidents. Of significant value in this regard

are investigation reports produced by the United States Chemical Safety and Hazard Investigation Board (U.S. Chemical Safety Board, or CSB). A recent analysis of these reports by Amyotte, MacDonald, and Khan [16] focused on the lessons to be learned with respect to inherently safer design and safety management systems.

- Interdisciplinary considerations are key to future successes in meeting industry's process safety research needs. The process industries have been moderately successful with the incorporation of chemistry as practiced by industrial chemists into the domain of chemical/process engineering. (Much, however, remains to be done in this regard; witness the continuing need for the greater participation of chemistry researchers in the field of inherently safer design.) Elements of other disciplines such as game theory and multi-attribute decision making have also been adopted by process safety researchers. What has not been as widely adopted in process safety research are some of the research findings from the social sciences. Significant in this regard is the work of sociologist Andrew Hopkins (e.g., Hopkins [17-20]).

The last bullet point above touches on an issue that is arguably at the heart of process safety research needs: *Are technical solutions enough, or are more people-centered considerations required?* Consider two of the slides (adapted) from the presentation by Hendershot et al. [11]:

- What next? Future challenges
  - Globalization of the chemical industry
    - Establishing safety culture in developing countries
  - Continued economic pressures—the need to do more with fewer resources
  - Maintaining a good process safety culture and management system at a time of frequent mergers, acquisitions, divestitures, and other business environment changes
  - Chemical process safety in industries other than the traditional chemical process industries (biotechnology, electronics, food, pharmaceuticals, etc.)
- What next? Future challenges
  - Complacency
    - Do some people think process safety is a problem that has been solved?
    - Will good experience threaten the programs that are responsible for that good experience?

- Education and awareness in broad industry community
- Examples are reactivity hazards, dust explosions

The terms *globalization, culture, economics, management, complacency, education, awareness* all represent concepts that are clearly important to industry. But are these concepts firmly captured within the skill set of process safety researchers with a primary background in science and engineering? This is suggested as food for thought in contemplating the process safety research needs of the twenty-first century.

12.2.2 Hydrogen-Specific Process Safety Research Needs

Is there reason to expect that process safety research needs specific to hydrogen are different from those in general? We would argue that the answer is no—at least at the macroscopic level. There will always be material hazard issues, such as those identified in Section 12.1, that a generic review (Section 12.2.1) cannot capture. The points raised in Section 12.2.1 are, however, consistent with the extensive analysis in previous chapters (e.g., Chapter 7 on inherently safer design and Chapter 8 on safety management systems).

The last columns in Tables 12.1 and 12.2 show the results of process safety keyword searches in the *International Journal of Hydrogen Energy*. It is encouraging to see some level of activity in most of the areas surveyed, albeit typically at a lower level than in the three process safety journals. Two thoughts emerge:

- Should the non-process-oriented hydrogen industries attempt to learn more from the process industries in terms of transferable research knowledge? (Witness, for example, the comment that hydrogen distribution systems including refueling stations should be built with *inherently safer* features [4].)
- Is it time for an archival, peer-reviewed journal devoted solely to hydrogen safety?

References

1. Guy, K.W.A., The hydrogen economy, *Process Safety and Environmental Protection*, 78 (4), 324–327, 2000.
2. Dahoe, A.E., and Molkov, V.V., On the implementation of an international curriculum on hydrogen safety engineering into higher education, *Journal of Loss Prevention in the Process Industries*, 21 (2), 222–224, 2008.

3. Salvi, O., Process Safety Research and its Impact on Sustainability and Resilience of the Society, Plenary Paper, A Frontiers of Research Workshop, Mary Kay O'Connor Process Safety Center, Texas A&M University, College Station, TX (October 21–22, 2011).
4. Pasman, H.J., and Rogers, W.J., Safety challenges in view of the upcoming hydrogen economy: An overview, *Journal of Loss Prevention in the Process Industries*, 23 (6), 697–704, 2010.
5. Jordan, T., Adams, P., Azkarate, I., Baraldi, D., Barthelemy, H., Bauwens, L., Bengaouer, A., Brennan, S., Carcassi, M., Dahoe, A., Eisenrich, N., Engebo, A., Funnemark, E., Gallego, E., Gavrikov, A., Haland, E., Hansen, A.M., Haugom, G.P., Hawksworth, S., Jedicke, O., Kessler, A., Kotchourko, A., Kumar, S., Langer, G., Stefan, L., Lelyakin, A., Makarov, D., Marangon, A., Markert, F., Middha, P., Molkov, V., Nilsen, S., Papanikolaou, E., Perrette, L., Reinecke, E.-A., Schmidtchen, U., Serre-Combe, P., Stocklin, M., Sully, A., Teodorczyk, A., Tigreat, D., Venetsanos, A., Verfondern, K., Versloot, N., Vetere, A., Wilms, M., and Zaretskiy, N., Achievements of the EC Network of Excellence HySafe, *International Journal of Hydrogen Energy*, 36 (3), 2656–2665, 2011.
6. Molkov, V., Hydrogen safety research: State-of-the-art, in Proceedings of the 5th International Seminar on Fire and Explosion Hazards, Edinburgh, UK (April 23–27, 2007).
7. Molkov, V., Preface. Special Issue on Hydrogen Safety, *Journal of Loss Prevention in the Process Industries*, 21 (2), 129–130, 2008.
8. Amyotte, P.R., Process Safety Research Needs from the Industry Perspective, Plenary Paper, A Frontiers of Research Workshop, Mary Kay O'Connor Process Safety Center, Texas A&M University, College Station, TX (October 21–22, 2011).
9. Amyotte, P.R., and Khan, F.I., *An Inherently Safer Approach to Dust Explosion Risk Reduction*, Strategic Project Grant (Safety and Security) No. 396398, Natural Sciences and Engineering Research Council of Canada, 2010.
10. Amyotte, P.R., *Are Classical Process Safety Concepts Relevant to Nanotechnology Applications?* Journal of Physics: Conference Series 304 (Nanosafe2010: International Conference on Safe Production and Use of Nanomaterials), 012071, 2011.
11. Hendershot, D.C., Ventrone, T.A., Schwab, R.F., Ormsby, R.W., Davenport, J.A., and Bradford, W.J., History of Process Safety and Loss Prevention in the American Institute of Chemical Engineers, Presented at the American Chemical Society, National Meeting, Washington, D.C. (August 28–September 1, 2005).
12. Kletz, T.A., The origin and history of loss prevention, *Process Safety and Environmental Protection*, 77 (3), 109–116, 1999.
13. Knegtering, B., and Pasman, H.J., Safety of the process industries in the 21st century: A changing need of process safety management for a changing industry, *Journal of Loss Prevention in the Process Industries*, 22 (2), 162–168, 2009.
14. Qi, R., Prem, K., Ng, D., Ranes, M., Yun, G., and Mannan, M.S., Challenges and needs for process safety in the new millenium, *Process Safety and Environmental Protection*, 90(2), 91–100, March 2012.
15. Kirchsteiger, C., Impact of accident precursors on risk estimates from accident databases, *Journal of Loss Prevention in the Process Industries*, 10 (3), 159–167, 1997.
16. Amyotte, P.R., MacDonald, D.K., and Khan. F.I., An analysis of CSB investigation reports concerning the hierarchy of controls, *Process Safety Progress*, 30 (3), 261–265, 2011.

17. Hopkins, A., *Lessons from Longford. The Esso Gas Plant Explosion*, CCH Australia Limited, Sydney, Australia, 2000.
18. Hopkins, A., *Safety, Culture and Risk. The Organisational Causes of Disasters*, CCH Australia Limited, Sydney, Australia, 2005.
19. Hopkins, A., *Failure to Learn. The BP Texas City Refinery Disaster*, CCH Australia Limited, Sydney, Australia, 2009.
20. Hopkins, A. (Editor), *Learning from High Reliability Organisations*, CCH Australia Limited, Sydney, Australia, 2009.

13

Legal Requirements for Hydrogen Safety

13.1 General Aspects and Definitions

Regulations, codes, and *directives* are legal requirements imposed by legislative bodies (parliaments, governments, the European Union) and are mandatory for everyone dealing with a certain activity. On the contrary, *standards, guidelines,* and *codes of practice* are voluntary documents useful for interested industrial organizations engaged in the relevant activity.

Codes of practice refer usually to best practices with regard to safe handling, operation, and maintenance of a product. *Guidelines* or *guides* are documents written for a specific organization or specific users and provide guidance to ensure best practice and safety, or for providing information and analyzing codes, standards, and regulations on the recommended way to meet these requirements. The *state of the art* is the most advanced technique available currently. *Best engineering practices* are the best available practices for design and construction, or for operation of machines and other devices in industry and other activities.

In the European Union, a *directive* is the legal requirement that fosters common ground in all European countries. Directives are compulsory only after a predetermined period of time after their adoption by the national legislation of every member country. Unfortunately, there is not yet a European directive for hydrogen. Matters concerning any substance not specifically covered by a directive are usually mentioned incidentally in other application-specific directives. As a result, only national legislation may define minimum levels for hazardous substances including hydrogen.

With reference to pressurized containers, hydrogen is covered by the European Pressure Equipment Directive (PED) and also by the Transportable Pressure Equipment Directive (TPED), which in turn refers to international agreements for safe transport of hazardous goods, such as ADR (road), RID (rail), IMO (sea), and ADNR (inland waterways). Permits to use motor vehicles internationally are obtained via the UN ECE (United Nations Economic Commission for Europe).

13.1.1 ATEX Directive

With regard to regulation on the prevention of damage from the release of flammable gases (including hydrogen), the directives 94/9/EC (known as ATEX 100 Directive) and 99/92/EC (known as ATEX 118 Directive) are relevant. ATEX derives its name from the French title of the 94/9/EC directive: *Appareils destinés à être utilisés en ATmosphères EXplosives*. Major accident prevention and mitigation of consequences to humans, installations, and the environment are covered by the Seveso II Directive.

ATEX Directive EC/1999/92 defines the minimum requirements for improving safety and health in a working environment by minimizing risk from a potential explosive atmosphere. In this sense, hydrogen poses a risk of forming an explosive atmosphere when mixed with air. To prevent such an event, the employer is obliged to take a series of measures based on the following principles:

- Prevention of formation of ATEX and, if this is not possible,
- Avoidance of ignition of ATEX, and
- Mitigation of explosion effects.

To accomplish this role, the employer has to perform risk assessment studies aimed at determining:

- The likelihood of an ATEX event.
- The likelihood of presence of active and effective ignition sources.
- The scale of anticipated effects in relation to installations, substances, and processes used.

The employer should also classify hazardous zones as follows.

- Zone 0: Places with continuous presence of ATEX, or for a long period of time, or with high frequency.
- Zone 1: Places with occasional risk of ATEX during normal operation.
- Zone 2: Places with unlikely risk of ATEX during normal operation and with short duration.

With regard to application of the ATEX Directive to hydrogen hazards, illustrative examples are the following:

- A process in which hydrogen is used (e.g., after ingress of air in a hydrogenation reactor).
- A closed space in which hydrogen may be produced via a metal powder reaction with water (especially when acidified).

- Hydrogen production during refueling of lead batteries.
- Release of hydrogen from pressurized facilities.

In addition to this ATEX Directive, many other directives are related to flammable gases including hydrogen, such as the Machinery Directive.

13.1.2 Other Official Documents

Considering the international standardization of hydrogen safety, the ISO Technical Committee 197 titled "Hydrogen Technologies" has published the following official documents:

- ISO 13984:1999 Liquid hydrogen - Land vehicle fuelling system interface
- ISO 14687:1999 Hydrogen fuel - Product specification
- ISO 14687:1999/Cor 1:2001 (Update of the above)
- ISO/PAS 15594:2004 Airport hydrogen fuelling facility operations
- ISO/TR 15916:2004 Basic considerations for the safety of hydrogen systems
- ISO 17268:2006 Compressed hydrogen surface vehicle refuelling connection devices.

In addition, committee ISO TC22 "Road Vehicles" has published the standards "Electrically propelled Road Vehicles" and SC 25 "Vehicles using gaseous fuels" through its subcommittee SC 21. The official hydrogen-related documents ISO 23273-2 "Fuel cell road vehicles - Safety specifications" and Part 2: "Protection against hydrogen hazards for vehicles fuelled with compressed hydrogen" have also been published by the committee SC21.

Some additional useful guidelines and other documents can be found in HySafe (Safety of Hydrogen as an Energy Carrier; see Chapter 9) [1] as listed below:

- Commission des Communautés Européennes, *Eléments pour un guide de sécurité hydrogène,* Expérimentations spécifiques, choix d'appareils et matériels adaptés, Volume 1, Rapport EUR 9689 FR, Luxembourg 1985.
- Commission des Communautés Européennes, *Eléments pour un guide de sécurité hydrogène, Aperçu d'ensemble,* Volume 2, Rapport EUR 9689 FR, Luxembourg 1985.
- FM Global, "Hydrogen," Property Loss Prevention Data Sheets 7-91, September 2000.
- IGC 15/96/E, *Gaseous Hydrogen Stations,* Industrial Gases Council, Brussels, Belgium.

- IGC 06/93/E, *Safety in Storage, Handling and Distribution of Liquid Hydrogen*, Industrial Gases Council, Brussels, Belgium.
- ISO/TR 15916, *Basic Considerations for the Safety of Hydrogen Systems* (*Considérations fondamentales pour la sécurité des systèmes à l'hydrogène*), first edition, 2004-02-15.
- American National Standard Institute, *Guide to Safety of Hydrogen and Hydrogen Systems*, American Institute of Aeronautics and Astronautics, ANSI/AIAA G-095-2004, Chap. 4. ANSI, Washington, D.C., 2004.
- NASA/TM—2003–212059, "Guide for Hydrogen Hazards Analysis on Components and Systems," Harold Beeson (Lyndon B. Johnson Space Center White Sands Test Facility), Stephen Woods (Honeywell Technology Solutions Inc. White Sands Test Facility); Published as TP-WSTF-937, October 2003.
- NASA, "NASA Glenn Safety Manual: Chapter 6 – Hydrogen," Revision Date: 9/03, Biannual Review.
- NFPA 50A, "Standard for Gaseous Hydrogen Systems at Consumer Sites," National Fire Protection Association, Quincy, Massachusetts, 1999.
- NFPA 50B, "Standard for liquefied hydrogen systems at consumer sites," National Fire Protection Association, Quincy, Massachusetts, 1999.
- NFPA 853, "Standard for the Installation of Stationary Fuel Cell Power Plants," National Fire Protection Association, Quincy, Massachusetts, 2003.
- NRCC 27406, "Safety Guide for Hydrogen," Hydrogen Safety Committee, National Research Council of Canada, Ottawa, 1987.

Certain guidelines and codes of practice are more substance oriented, among which are the American National Standard ANSI/AIAA G-095-2004, *Guide to Safety of Hydrogen and Hydrogen Systems* [2] of the American Institute of Aeronautics and Astronautics, the Final Report on *Guidelines for Use of Hydrogen Fuel in Commercial Vehicles* of the U.S. Department of Transportation [3] and the Code of Practice (COD) on *Safety in Storage, Handling and Distribution of Liquid Hydrogen* of the European Industrial Gases Association (EIGA) [4]. These are described in some detail in the following sections.

13.2 Hydrogen Facilities

The following are general guidelines to ensure safety in hydrogen storage and transfer areas. Good illumination, lightning protection, alarm systems, and gas detection and sampling systems should be provided in such a facility. More detailed safety guidelines concerning safety policy, safety in construction, operation, maintenance, and final disposition of a hydrogen facility, as well as safety measures in buildings and test chambers used in hydrogen service and emergency procedures can be found in the American National Standard ANSI/AIAA G-095-2004 [2] from which some synoptic information is given in this section, along with other references [5].

13.2.1 Electrical Considerations

Areas where flammable hydrogen mixtures are expected to occur or areas where hydrogen is stored, transferred, or used and where the hydrogen normally is contained are classified as highly hazardous zones according to internationally accepted regulations. All electrical sources of ignition should be prohibited in these areas, using approved explosion-proof equipment or selecting non-arcing approved equipment.

Explosion-proof equipment has an enclosure strong enough to contain the pressure produced by igniting a flammable mixture inside the enclosure. Since it is not gas tight, the joints and threads must be tight enough and long enough to prevent issuance of flames or gases that would be hot enough to ignite a surrounding flammable mixture.

Another method to prevent a gas explosion is to locate the equipment in an enclosure purged and maintained above ambient pressure with an inert gas. Intrinsically safe installations used in the facility should be approved for hydrogen service.

It is acceptable to use general-purpose equipment in general-purpose housing, if it is continuously purged with clean air or nitrogen. In this case, no hydrogen sources must be plumbed into the equipment and positive indication of continued purge must be provided. By putting items that might become ignition sources outside the hazardous area, the cost of an installation will be reduced and safety increased. Equipment installed in a hazardous area, but not required during hazardous periods, may be built with general-purpose equipment provided it is disconnected before the hazardous period begins. The conduits for such systems must be sealed when they leave the hazardous area.

13.2.2 Bonding and Grounding

Mobile hydrogen supply units must be electrically bonded to the system before discharging hydrogen according to regulations.

Liquefied hydrogen containers, static and mobile, and associated piping must be electrically bonded and grounded using proper sizes of grounding conductors and acceptable connections based on the expected amperage to ground.

All *off-loading facilities* must provide easily accessible grounding connections and be located outside the immediate transfer area. Facility grounding connections should have resistances less than 10 ohm. Transfer subsystem components should be grounded before subsystems are connected.

13.2.3 Hydrogen Transmission Lines

Hydrogen transmission lines carrying hydrogen from trailers and storage vessels must be aboveground installations. Lines crossing roadways should be installed in concrete channels covered with an open grating. Hydrogen lines crossing roadways should be installed in concrete channels covered with an open grating and should not be located beneath electric power lines. Area surfaces located below liquefied hydrogen lines from which condensed liquid air may drop must be constructed of noncombustible materials such as concrete. Asphalt must not be used.

Piping leaks have been the initiating event in many hydrogen and other gaseous fuel accidents. A hydrogen leak, due to its very low density, can easily find the way to upper floors in a building, traveling through connecting ducts and other openings, and cause secondary explosions in other spaces.

13.2.4 Elimination of Ignition Sources

Installations using hydrogen should be protected from *lightning* by lightning rods, aerial cable, and ground rods suitably connected. Lightning strikes may cause inducing sparks; therefore, all equipment in a building should be bonded and grounded to prevent sparks.

Static electricity may be generated in moving machinery belts or in flowing fluids containing solid or liquid particles. The measures taken to limit electrostatic charge generation and accumulation include:

- Bonding and grounding of all metal parts within a system
- Use of conductive machinery belts
- Personnel clothes made of antistatic fibers
- Conductive and nonsparking floors

Sparks may also be generated by other mechanisms such as *friction and impact*. Even spark-proof tools can cause ignitions because the energy required for ignition of flammable hydrogen-air mixtures is extremely small. So, spark-proof tools should be used and in addition they should be used with caution to prevent slipping, glancing blows, or dropping, all of which can cause sparks. Extra care should be taken if spark-proof tools are not available.

13.2.5 Hot Objects, Flames, and Flame Arrestors

The measures taken to eliminate ignition by flames and hot objects and prevent proliferation of a fire to other areas (domino effect) include prohibition of open flames, welding, or cutting within the exclusion area around a hydrogen facility, equipment of internal combustion systems with exhaust system spark arrestors, and carburetor flame arrestors. Only *flame arrestors* specifically designed for hydrogen applications must be used taking into account the oxidant present. Flame arrestors can quench a flame on the basis of sufficient heat removal from the gas mixture. It is quite difficult to develop flame arrestors and explosion-proof equipment for hydrogen due to its small quenching distance (0.6 mm). Sintered-bronze flame arrestors may be effective in stopping hydrogen flames, whereas sintered stainless steel is not as effective. Arrestors should be well maintained to minimize accidental ignition because many accidents have been caused by poor maintenance of safety devices.

13.2.6 Design and Construction of Buildings

Buildings in which hydrogen is used must be constructed of light and non-combustible materials on a sufficiently strong frame. Windowpanes must be made of shatterproof glass or plastic. Floors, walls, and ceilings should be designed and installed to limit the generation and accumulation of static electricity and have a fire resistance rating of at least 2 h.

13.2.6.1 Explosion Venting

Explosion venting must be provided only in the exterior walls or the roof. The venting area should not be <0.11 m^2 per cubic meter of room volume. Vents, designed to relieve at a maximum internal pressure of 1.2 kPa, may consist of one or a combination of walls of light material, lightly fastened hatch covers, or outward-opening swinging doors in exterior walls or roof.

Doors should be hinged to swing outward in an explosion and must be readily accessible to personnel. *Walls* or partitions must be continuous from floor to ceiling and securely anchored. At least one wall must be an exterior wall, and the room must not be open to other parts of the building. Only indirect means, such as steam and hot water, should be used for *heating* in rooms containing hydrogen.

13.2.6.2 Test Facilities

Test facilities (chambers, cells, or stands) should be so constructed as to adhere to appropriate safety standards and guidelines. Test cells that cannot be ventilated sufficiently to prevent explosive hazards should be provided with an inert atmosphere of nitrogen, carbon dioxide, helium, steam, or other inert gas. The test cell pressure should be higher than atmospheric to avoid inflow of air. Nevertheless, the system design must prevent asphyxiation of personnel in adjacent areas. The system design must prevent the entrance of personnel in the cell unless confined space conditions are safe.

A partial vacuum may be used to restrict oxidants (e.g., oxygen) in a test chamber. In this case vacuum should be sufficient to limit the pressure of an explosion to a value that the system can withstand. Reflected shock waves should be taken into consideration in the design of heads, baffles, and other obstructions in a pipe run. Ultimate stress values should be used because of the violent nature of an explosion.

13.2.7 Placarding in Exclusion and Control Areas

Exclusion areas must have placards, postings, and labels displayed, so personnel are made aware of the potential hazard in the area.

Gaseous hydrogen systems must be permanently placarded as follows:

HYDROGEN – FLAMMABLE GAS - NO SMOKING - NO OPEN FLAMES.

Storage sites for liquefied hydrogen systems must be fenced and posted to prevent entrance by unauthorized personnel and placarded as follows:

LIQUEFIED HYDROGEN – FLAMMABLE GAS - NO SMOKING - NO OPEN FLAMES

A sign must be placed on the container near the pressure-relief valve vent stack or on the vent stack warning against spraying water on or in the vent opening.

The maximum number of workers and transients permitted at any time and the maximum amount of propellant materials and their Groups/Classes must be placarded and posted in a conspicuous place in all buildings, cells, rooms, and storage areas containing liquefied hydrogen. Safety showers must be placarded:

NOT TO BE USED FOR TREATMENT OF CRYOGENIC BURNS.

13.2.8 Protection of Hydrogen Systems and Surroundings

13.2.8.1 Barricades

Barricades are constructed to protect uncontrolled areas from the effects of a hydrogen system failure and to protect a hydrogen system from the hazards of adjacent or nearby operations. Pressure vessels, piping, and components are designed in a way that failure caused by overpressure or material deficiencies will not produce shrapnel. Barricades should be constructed adjacent to the expected fragment source and in a direct line of sight between it and the facility to be protected because barricades have been shown to be most effective against fragments and only marginally effective in reducing overpressures at extended distances from them [6].

The *housing of equipment* provides partial protection in many cases. Shrapnel protection can be achieved by blast curtains or blast mats placed adjacent to the equipment to be protected.

Barricades are commonly constructed as mounds and single-revetted barricades. A mound (earthworks) is an elevation of sloped dirt with a crest at least 0.91 m wide, whereas single-revetted barricades are mounds supported on a retaining wall on the side facing the hazard source.

Simulations [6] and experimental work [7] have shown that:

- Barricades should be designed to block the line of sight between equipment from which fragments can originate and the protected items.
- Barricades should be placed adjacent to the fragment source for maximum protection.
- The efficiency of barricades depends on the height aboveground and barricade location, dimensions, and configuration.
- Barricades can reduce peak overpressures and impulses behind the barricades, but reflections on obstacles may amplify blast waves behind barricades.
- Single-revetted barricades are more efficient than mounds.

13.2.8.2 Liquid Spills and Vapor Cloud Dispersion

A liquid H_2 (LH_2) spill from a storage vessel will result in a brief period of ground-level flammable cloud travel. The quick evaporation of the liquid causes the hydrogen vapors to mix quickly with air, dilute to nonflammable concentrations, warm up, and become positively buoyant. Accident prevention measures in such cases would be either natural dispersion or confinement of the spill [8].

If barricades are chosen, these should not excessively confine the vapor cloud formed because this may lead to detonation rather than simple burning of escaped hydrogen. LH_2 spills in an open-ended (U-shaped) bunker may produce detonation of the hydrogen-air mixture even without a roof [2].

The use of dikes and barricades around storage tanks is not recommended for LH_2, although this is required for liquid natural gas (LNG) because it may prolong evaporation and ground-level travel of the flammable cloud. Hydrogen detectors should be positioned to indicate the possible ground-level travel of flammable mixtures. Sewer drains must not be located in an area in which a liquefied hydrogen spill could occur.

13.2.8.3 Shields and Impoundment Areas

Impoundment areas and shields to control the extent of liquid and vapor travel caused by spills should be included in the design of a facility using hydrogen. The loading areas and the terrain below the transfer piping should be directed to a sump or impoundment area. Crushed stone should be used in the impoundment area to provide added surface area for liquefied hydrogen dissipation.

Ignition of hydrogen-air mixtures in free space usually results in combustion or deflagration, but with a certain degree of confinement a deflagration can evolve into a detonation (termed DDT). Consequently, careful design of installations should eliminate possible confinement developed by the equipment or buildings because initial combustion or deflagration of hydrogen-air mixtures may evolve to DDT.

13.2.9 Quantity-Distance Relationships

In spite of the numerous efforts to determine quantity-distance relationships between LH_2 storage facilities and inhabited buildings in various countries, the results are largely uncertain because of the different assumptions made by each institution. The International Atomic Energy Agency [9] has taken up many of these quantity-distance relationships as specified in codes and standards of different countries and institutions. They are shown for comparison in Figure 13.1.

Many efforts to define a quantity-distance relationship have turned to the so-called *cube-root scaling law*. This law relates the mass of a flammable substance with distance to define safety distance and is expressed by the simple equation:

$$R = k * M^{1/3} \tag{13.1}$$

where R is the distance from a mass M (m) of a flammable substance (kg).

The factor k depends on the type of building to be protected. According to German recommendations this factor takes the values 2.5 to 8 for working buildings, 22 for residential buildings, and 200 for supposed no damage. The factor k may be modified by damping parameters if protective measures such as earth coverage or protective walls are taken.

13.3 Hydrogen Fueling of Vehicles

The Final Report on *Guidelines for Use of Hydrogen Fuel in Commercial Vehicles* of the U.S. Department of Transportation [3] is a thorough text covering this

1 US Dep. Defense Instruction No. 4145.21 (1964)
2 National Fire Protection Association (NFPA), Boston (1973)
3 US Army Material Command Safety Manual No. 385-224 (1964)
4 Bureau of Mines, Pittsburgh (1961)
5 German Fed. Ministry of the Interior, Bonn (1974) (nuclear power plants) for liquefied gases
6 High-Pressure Gas Control Legislation, Japan (category I)

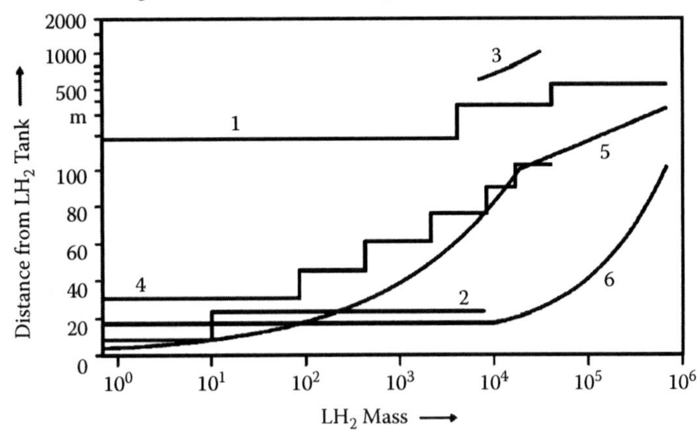

FIGURE 13.1
Safety distances of LH_2 storage tanks from inhabited buildings as a function of LH_2 mass in various countries and institutions (please note scale change on the ordinate). (From International Atomic Energy Agency, *Hydrogen as an Energy Carrier and Its Production by Nuclear Power*, IAEA-TECDOC-1085, 1999. http://www.iaea.org/inisnkm/nkm/aws/htgr/fulltext/30027279.pdf.)

issue in detail. Thus the following sections make use of its suggestions in brief, and it is recommended that those dealing with hydrogen safety with regard to use as a vehicular fuel consult this valuable report.

13.3.1 Guidelines for Hydrogen Systems on Vehicles

13.3.1.1 Compressed Hydrogen Systems

Vehicle design—For safety reasons hydrogen-fueled vehicles should be properly marked with diamond-shaped labels indicating "Compressed Hydrogen." The labels should be legible from 15 m in daylight. The gas pressure should be reduced gradually in three stages: fuel storage system (up to 345 bar), motive pressure circuit (up to 12 bar), and low-pressure circuit (up to 1 bar). The whole system should be equipped with proper pressure relief devices, isolation valves, and flow regulators.

A safety factor of three is recommended for the whole gas system according to NFPA 2005 [10], at least for vehicles circulating or constructed in the United States.

Furthermore, many other recommendations should be followed, some of which are listed below:

- The fuel cylinders should be permanently marked "Hydrogen" and securely mounted and protected from damage by road debris.
- All cylinders should be equipped with PRD (pressure relief devices), TRD (thermal relief devices) and manual shut-off valves.
- Isolation of the cylinders from the rest of the system should be done with electrically activated "fail safe" valves.
- Special care should be taken in the selection of construction materials in contact with hydrogen to ensure they are not subject to hydrogen embrittlement.
- Not only the fuel system, but also the whole engine system should be electrically bonded and finally grounded to avoid the buildup of static electricity.
- All vehicle compartments should be efficiently ventilated to avoid accumulation of hydrogen in case of a leak exceeding a hydrogen concentration of 1 percent by volume (25 percent of the lower flammable limit).
- Hydrogen sensors should be installed on the vehicle and be properly connected to the vehicle control system to activate an alarm and automatic system shutdown when hydrogen concentration exceeds 1 to 2 percent by volume.
- An automatic system shutdown should be installed such that it is triggered by detection of leaked hydrogen, excess fuel flow, vehicle crash, or other system default.
- The vehicle operator should have easy accessibility to an on/off switch allowing shutting down the fuel cell system, disconnecting the traction power, deenergizing high-voltage equipment, and shutting off fuel supply.
- An interlock to the vehicle fueling port should prevent fueling unless the fuel cell system is shut down and the traction system is off.
- Electrical bonding of the fueling receptacle to the vehicle chassis is a must, accompanied by efficient grounding.
- When needed, hydrogen venting should be done by an outlet at the top of the vehicle.

Operation and maintenance—All *operators* of compressed-hydrogen-fueled vehicles should receive special training on hydrogen hazards and emergency response. Those dealing with *maintenance* should never use noncertified replacement parts. Parts destined for natural gas should not be used in hydrogen systems. Special caution is needed for checking any probable leak. Furthermore:

- The maintenance technicians should never loosen any joints under hydrogen pressure or over-tighten joints above the levels specified by manufacturers.
- If exposed to the atmosphere, purging of any component, and especially of hydrogen fuel cylinders, is only allowed with nitrogen before refilling.
- Thorough visual inspection of hydrogen cylinders for nicks, dents, and cuts must be conducted at least every 36 months or 58,000 km.
- Periodic checks and calibrations should always be performed according to the manufacturer's service manual.
- Warning lights and alarms should not be ignored and automatic system shutdown should not be overridden.
- No repairing of fuel lines is allowed, only replacement.
- Isolation of the fuel system, disconnection of the battery, and switching off of the main switch are mandatory before servicing the vehicle.
- No smoking or use of cell phones is allowed when servicing the vehicle.
- The internal pressure in hydrogen cylinders should always remain above atmospheric pressure; if not, purging with nitrogen is indispensable before refilling with hydrogen.

13.3.1.2 Liquid Hydrogen Systems

Experience for cryogenic systems onboard is not as common as for compressed gases. In addition, special care should be taken with respect to hazards stemming from the extremely low temperatures.

Vehicle design—Labels should be put on the exterior of the vehicle saying "Liquid Hydrogen," which should be legible from a distance of at least 15 m. The cryogenic tanks should also be marked with the word "hydrogen," securely fastened, routed away from heat sources onboard, and protected from road debris. In the absence of certification standards for liquid hydrogen tanks, at least the existing standards for LNG should be applied for liquid hydrogen (including drop and flame tests).

Furthermore, many other recommendations should be followed, some of which are listed below:

- The cryotanks should be equipped with a pressure relief valve, whose outlet should empty into a hydrogen diffuser. The latter should be able to dilute hydrogen with air to a concentration less than 1 percent by volume.
- A manual shutoff valve should be installed on the vehicle to isolate the hydrogen tank from the rest of the fuel system.

- A liquid level gauge should be installed on each cryotank, with indication in the driver's cab and a pressure gauge readable near the tank.
- Each cryotank should be equipped with at least one electrically activated valve isolating the tank from the rest of the fuel system. These valves should be of the "fail safe" type.
- Fuel lines should not pass through the passenger compartment.
- All materials coming in contact with liquid hydrogen should be chosen to withstand the liquid hydrogen temperatures, and the materials coming in contact with gaseous hydrogen should be resistant to hydrogen embrittlement.
- The fuel and engine system of a liquid hydrogen vehicle should be electrically bonded and finally grounded to prevent buildup of static electricity.
- A pressure relief valve should be installed on every liquid hydrogen line isolated between two potentially closed valves to deal with venting hydrogen as it vaporizes because of line heat-up.
- Hydrogen sensors should be installed on the vehicle, providing an alarm and automatic system shutdown if the hydrogen concentration exceeds 1 to 2 percent by volume.
- Excess flow valves should also be installed on the vehicle capable of stopping fuel flow when the flow exceeds the set threshold.
- The liquid hydrogen vehicle should be equipped with an inertial crash sensor to automatically shut down the vehicle in case of a crash. There should be the possibility to allow the vehicle to operate for a short time with the aid of a switch in an emergency, such as moving out of high-speed traffic or off a railroad track.
- The vehicle operator should have easy accessibility to an on/off switch allowing shutting down the fuel cell system, disconnecting the traction power, deenergizing high-voltage equipment, and shutting off the fuel supply.
- An interlock to the vehicle fueling port should prevent fueling unless the fuel cell system is shut down and the traction system is off.
- When needed, hydrogen venting should be done by an outlet at the top of the vehicle.

Operation and maintenance—All *operators* of LH2-fueled vehicles should receive special training on hydrogen hazards and emergency response. Those dealing with *maintenance* should never use noncertified replacement parts. Parts destined for natural gas should not be used in hydrogen systems. Special caution is needed for checking any probable leak. Lines transmitting liquid hydrogen must be thermally insulated, with the outer layer of insulation vapor sealed. All components coming in contact with liquid hydrogen must withstand the extremely low temperatures.

Furthermore, many other recommendations relating not only to high pressure but also to the extremely low temperatures encountered with liquid hydrogen should be followed, some of which are listed below:

- Maintenance technicians should never work on a liquid hydrogen system without wearing appropriate personal protective equipment (e.g., safety glasses or full-face shield, leather gloves and boots, long-sleeved shirt, long pants without cuffs outside of the boots).
- They should never loosen or over-tighten joints connected to hydrogen lines under pressure or containing liquid hydrogen.
- They should never disturb the insulation of liquid hydrogen lines or cryotanks still containing liquid hydrogen.
- They should never allow air to enter any component of the hydrogen system. If that happens accidentally (because of lower than ambient pressure in a cryotank) helium should be used to purge before refilling. Nitrogen is not allowed to be used in liquid hydrogen systems because it will liquefy and then solidify when in contact with liquid hydrogen, thus plugging lines and valves.
- Connections of the whole hydrogen system should be checked periodically according to the manufacturer's service manual.
- The outer surface of fuel lines and cylinders should be checked for any damage, as well as the exterior insulation layer of the cryotank for any damage causing loss of its function as a vapor barrier.
- The technicians should periodically check the hydrogen sensors and the hydrogen diffuser fan according to the manufacturer's service manual.
- Warning lights and alarms should not be ignored and the automatic system shutdown should not be overridden.
- No repairing of fuel lines is allowed, only replacement.
- Isolation of the fuel system, disconnection of the battery, and switching off of the main switch are mandatory before servicing the vehicle.
- No smoking or use of cell phones is allowed when servicing the vehicle.
- Before fueling, the vehicle should be bonded and grounded. The ullage space specified by the manufacturer should be maintained and no overfilling is allowed.

13.3.1.3 Liquid Fuel Reformers Onboard

In an effort to minimize the load of a hazardous material in a process according to the principles of inherent safety discussed in Chapter 7, *fuel reformers* have been developed, which reform a liquid fuel such as diesel fuel, gasoline, or methanol to a *reformate* containing 45 to 75 percent hydrogen. The other constituents are carbon dioxide, nitrogen, water, and sometimes carbon monoxide.

The reformate is produced on demand and can be used to feed a hydrogen fuel cell, thus eliminating the need to carry significant quantities of either gaseous or liquid hydrogen on the vehicle. After shutting down the engine, the remaining reformate is usually vented and no hydrogen is left onboard.

Vehicle design—Labels should be put on the exterior of the vehicle saying "Liquid Hydrogen," which should be legible from a distance of at least 15 m. When the reformer is in operation, the temperature of the fluids inside it may be between 93 and 816°C. This is why the whole system should be housed in a shielded package to prevent operators and technicians from coming in contact with hot surfaces.

Although the quantities of hydrogen onboard are minimal, sensors should be installed on the vehicle, providing an alarm and automatic shutdown when hydrogen or carbon monoxide concentrations in any compartment exceed a preset threshold. Remaining reformate in any system should be vented upon shutdown.

Operation and maintenance—All operators and maintainers of liquid fuel reformer vehicles should receive special training on hydrogen hazards and emergency response. This training should include information on the high temperatures encountered in leaking fluids from the reformer, which can cause serious burns.

The reformers should be designed as "line replaceable units" in a way that the mechanic will not open the hot box to make repairs. Line connections should be periodically checked for leaks according to the manufacturer's service manual. Any air and fuel filters, as well as desulfurization traps, must periodically be replaced accordingly.

13.3.2 Guidelines for Hydrogen Fueling Facilities

Fueling stations are supposed to be designed, constructed, and operated taking into account the hazardous properties of hydrogen and according to current codes, standards, and regulations in every country. The following information addresses mainly vehicle operators.

13.3.2.1 Compressed Hydrogen Fueling

The experience from compressed natural gas fuel dispensers is indispensable to the development of compressed hydrogen fuel dispensers because both are very similar. The main difference is the supply of the fuel gas; natural gas fuel stations are usually supplied with gas from pipelines, whereas hydrogen gas fuel stations are usually supplied with tube trailers hauled by trucks (Figure 13.2).

FIGURE 13.2
Hydrogen tube trailer (reference from unclassified U.S. DOT report [3]). (Photo courtesy of Sunline Transit)

The fueling system for compressed hydrogen is comprised of a coupling on the fueling nozzle that mates with a compatible coupling on the vehicle. These are locked together when the attendant turns a lever (Figure 13.3).

When liquid hydrogen is supplied to the fueling station a vaporizer needs to be installed as well. This is essentially a heat exchanger to vaporize the liquid hydrogen and raise the temperature of the resulting gas to ambient temperature if hydrogen has to be dispersed in a gaseous state (Figure 13.4).

FIGURE 13.3
Compressed hydrogen fuel nozzle (reference from unclassified U.S. DOT report [3]). (Photo courtesy of Santa Clara Valley Transportation Authority)

FIGURE 13.4
Liquid hydrogen storage tank and vaporizer (reference from unclassified U.S. DOT report [3]). (Photo courtesy of Santa Clara Valley Transportation Authority)

In some cases hydrogen is produced on-site by reforming natural gas, as occurs with onboard fuel reformers, but in this case the reformer is much larger and a separator has to be installed for the separation of hydrogen from the other ingredients of the reformate.

Design—As is the case with all flammable gases and vapors, hydrogen fuel stations are located outdoors in the open air so that any gases released can be easily and safely dispersed in the air. Any construction should be designed in a way that no collection of escaping gas is allowed. Thus, if, for instance, there is a canopy, this should be constructed with upward slopes (Figure 13.5).

Generally, measures should be taken to protect the station from any damage caused by vehicles, the public, and other nearby activities. According to the NFPA [10], the minimum distance of any component of the fueling station from other buildings or public roadways should be 3 m.

Other recommendations are:

- Measures with special devices that shut off fuel flow should also be taken to prevent accidental or intentional movement of a vehicle with the hose connected.

FIGURE 13.5
Compressed hydrogen fueling station (reference from unclassified U.S. DOT report [3]). (Photo courtesy of Alameda Contra-Costa Transit District)

- Explosion-proof electrical equipment must always be installed and all components including the fueling hose must be bonded and grounded. The vehicle must also be bonded and grounded, usually via the fueling hose.
- Open-flame heaters are not allowed at the fueling station; hot air, steam, or hot water should be used for indirect heating instead.
- The fueling system should have provisions for controlling the final pressure so as not to exceed the rated pressure, and for not permitting refilling if the pressure inside the vehicle fuel tanks is less than normal atmospheric pressure. The latter case indicates that air might have entered the tank and so there is a possibility of an explosive mixture having been formed inside the tank.
- All hydrogen storage tanks should be equipped with a PRD (pressure relief device)/TRD (thermal relief device). Any hydrogen venting should be routed to an outlet higher than any surrounding structure, which should be equipped with a hydrogen diffuser or flame arrestor.
- If a compressed hydrogen fuel station also has a liquid hydrogen storage tank (cryotank), this should be equipped with a PRD, as well as any line isolated between two closed valves.
- The equipment at the station should include emergency stop systems and hydrogen sensors, including ultraviolet flame sensors to

detect flames in the vicinity of the fuel dispenser. Automatic shutdown should occur when the hydrogen concentration in the air is more than 1 percent by volume.

- Dry powder fire extinguishers should be available at the hydrogen station for fire fighting.
- Signs at the fuel station should remind customers of the proper behavior in the area of the station, as well the appropriate actions in case of an emergency.

Operation and maintenance—All personnel at a hydrogen fueling station should be trained in hydrogen safety and particularly the operational characteristics of these stations. All activities that might result in the creation of sparks, flames, or hot points, such as smoking, use of cell phones, welding or cutting metals, must be forbidden in the area.

A vehicle should be allowed to be fueled only after the driver has turned off the ignition and immobilized the vehicle. In addition to bonding and grounding via the fuel nozzle, additional grounding by a ground bond strap is recommended. Vehicle maintenance should not be allowed at a hydrogen fuel station and any other maintenance of the station must be performed only with nonsparking tools. At least annual maintenance of emergency equipment, such as hydrogen and flame sensors, should be performed according to the manufacturer's service manual [10].

13.3.2.2 Liquid Hydrogen Fueling

The main difference between compressed gaseous hydrogen fueling stations and liquid hydrogen fueling stations is that compressors and pumps are not used to pump out liquid hydrogen from cryotanks. This is accomplished by the addition of heat to the tank by a suitable safe heater. This energy input increases the pressure inside the tank, thus forcing out liquid hydrogen.

Liquid hydrogen fueling nozzles are more complicated than those used for compressed hydrogen fueling. In this case, the so-called male coupling is found on the fueling nozzle and the so-called female coupling is on the vehicle (Figure 13.6). Both are locked together with the aid of a lever. Only then can fueling begin.

Design—All safety measures applied to compressed gaseous hydrogen fueling stations are also applicable to liquid hydrogen fueling stations. Nevertheless, additional safety measures should be taken owing to cryogenic conditions found in this case. Such additional measures are:

- According to the NFPA [10], hydrogen cryotanks should not be larger than 600 gallons (2.271 m^3) and liquid hydrogen fueling dispensers should be located in the open air.

FIGURE 13.6
Liquid Hydrogen Fueling Nozzle and Vehicle Fuel Port (reference from unclassified U.S. DOT report [3]) (Photo courtesy of Air Products and Chemicals).

- The specified minimum set-back distances between liquid hydrogen storage tanks and occupied buildings or public roadways varies between 3 m and 22.8 m, depending on the size of the storage cryotanks. The minimum distance of liquid hydrogen dispensers from occupied buildings should be at least 7.6 m.
- Special note should be taken of the fact that a liquid hydrogen leak could find its way to drains or underground facilities, thus creating a potential explosion hazard after vaporization and formation of an explosive mixture with the air. Consequently, measures such as dikes around the cryotanks should be taken to prohibit this possibility and, in any case, give sufficient time for liquid hydrogen to vaporize before reaching underground facilities.
- Good insulation and an unimpaired outer surface of the insulation are indispensable for safety. Liquid hydrogen leaks are able to liquefy air that is oxygen enriched. Thus, liquefied air can pose a severe fire hazard if storage tanks, liquid hydrogen lines, and dispensers are mounted on combustible materials such as asphalt. In general, a concrete pad is recommended to prevent this hazard.
- Sometimes ice and frost are created due to the extremely low temperature of liquid hydrogen; such ice and frost formations should be cleaned up for safety reasons using a high-pressure air, nitrogen, or helium supply.

Operation and maintenance—In addition to the safety measures recommended previously for compressed hydrogen fueling stations, the following measures are required for liquid hydrogen fueling stations:

- The extremely low temperature of liquid hydrogen requires special care; otherwise, it will cause severe frostbite to those exposed to this hazard unprotected. Thus, persons connecting and disconnecting the liquid hydrogen nozzle from the vehicle should wear safety glasses and a full face shield, loose fitting insulated or leather gloves, leather boots at ankle height or higher, a long-sleeved shirt, and long pants without cuffs. Pant legs should be worn outside of the boots.

- Since ice and frost may be formed on the nozzle the attendant should check the mating surfaces before fueling. If this happens, the attendant may use high-pressure air, nitrogen, or helium to remove this ice and frost to prevent eventual leaking during fueling.
- Maintenance (at least every six months) of emergency equipment such as hydrogen and flame sensors should be performed according to the manufacturer's service manual [10].

13.4 Storage, Handling, and Distribution of Liquid Hydrogen

In 2002 the European Industrial Gases Association (EIGA) prepared a Code of Practice (COD) on safety in the storage, handling, and distribution of liquid hydrogen [4]. This COD deals with many issues concerning hydrogen safety, including:

- Physical, chemical, and biological properties of hydrogen.
- Customer installations (layout and design features, access to installations, testing and commissioning, decommissioning and removal of tanks, operations and maintenance, customer information).
- Transport and distribution of liquid hydrogen (road transport, tank container, transport by railway, waterways, and sea).
- Training and protection of personnel (gas supplier and customer, work permits).

The first two of these items are generally in accordance with the detailed information and recommendations given in Sections 13.1 and 13.2, which are based on the American National Standard ANSI/AIAA G-095-2004, *Guide to Safety of Hydrogen and Hydrogen Systems* [2] and on the *Guidelines for Use of Hydrogen Fuel in Commercial Vehicles* of the U.S. Department of Transportation [3]. However, the last two items give additional information and recommendations addressed to those dealing with the transport of liquid hydrogen by all transport means (with the exception of pipelines and portable containers such as pallet tanks and cylinders). An overview of the EIGA's COD follows.

TABLE 13.1

Selected Recommended Minimum Safety Distances for Liquid Hydrogen

	Items	Distance (m)
1	Technical and unoccupied buildings	10
2	Occupied buildings	20
3	Other LH_2 fixed storage	1.5
4	Other LH_2 tanker	3
5	Flammable gas storage	8
6	Open flame, smoking, welding	10
7	Public establishments	60
8	Railroads, roads, property boundaries	10
9	Overhead power lines	10

Source: *Code of Practice on Safety in Storage, Handling and Distribution of Liquid Hydrogen*, European Industrial Gases Association, 2002. With permission.

13.4.1 Customer Installations

Concerning customer installations, minimum safety distances are recommended depending on the endangered object or item. Safe distances are based on hydrogen cloud dispersion models, heat flux effects of a hydrogen flame, and local overpressure due to flame ignition as measured from the release point. Some of these recommended safety distances are shown in Table 13.1.

These safety distances can be reduced if certain protection measures are taken (such as a water spray curtain) between the liquid hydrogen installation and the exposure area. Nevertheless, such protection measures can be applied only for certain cases, such as items 1, 2, 5, and 8 in Table 13.1.

Provisional measures, in addition to those suggested in the *Guidelines for Use of Hydrogen Fuel in Commercial Vehicles* of the U.S. Department of Transportation [3], such as the construction of hydrogen storage installations in the open air, include barriers or bollards to eliminate vehicular impact. The hydrogen transfer should only take place within the user's premises. In an emergency, at least two separate outward-opening exits, at least 0.8 m wide, should be provided in technical buildings of the hydrogen installation.

The installation and operation of electrical systems in hydrogen installations (inside the distance given in Table 13.1, item 15) need to be in accordance with national regulations, standards, and codes of practice, and especially with the last amendment of Directive 79/196/EEC (electrical equipment for use in potentially explosive atmospheres). Only explosion-proof electrical equipment must be used inside the safety distances given in Table 13.1.

Bonding and grounding of all components is a necessity to preclude development of static electricity, especially when mechanical abrasion occurs or when a fluid containing particles flows past the surface of a solid. Most synthetic materials can very easily generate static electricity charges. The resulting electric sparks are usually sufficiently powerful to ignite the highly sensitive hydrogen.

All hydrogen vents need to be connected to a vent stack arranged to discharge in a safe place in the open air. The height of the vent stack outlet should be either 7 m above ground level or 3 m above the top of the tank, whichever is greater.

The EIGA code of practice also gives some recommendations on the rare occasion that gaseous hydrogen pipelines have to run in the same duct or trench with electrical cables. In this case all joints of the hydrogen pipelines need to be welded or brazed and these pipelines located higher than other pipelines.

Care should also be taken in the use of certain detection instruments that are not normally compatible with safety precautions required for hydrogen, such as gas chromatographs and flame ionization detectors.

13.4.2 Transport and Distribution of Liquid Hydrogen

13.4.2.1 Road Transport

According to the EIGA, all means of transport of dangerous goods by road should follow the European Agreement Concerning the International Carriage of Dangerous Goods by Road (ADR [**A**ccord Européen Relatif au Transport International des Marchandises **D**angereuses par **R**oute]). The ADR, under Council Directive 94/55/EC, harmonizes the law across the European Union. A consolidated "restructured" edition of ADR was published in 2005. The latest ADR rules apply from January 1, 2009.

The COD of the EIGA covers all operations, from the departure of the vehicle from the filling plant until it has completed all deliveries given in the route plan. These operations include route planning, periodic checking, parking, breakdown, product transfer, emergency procedures, and driver training.

Route planning. Planning needs to be very detailed, describing the exact itinerary the tanker or tank container will follow. In general, motorways and trunk roads should be preferred and tunnels should be avoided, as well as densely populated areas. Any diversion from the planned route should be made known to the home base as soon as it is safe to do so.

Periodic checking. The vehicle should be fully checked at the departure site and also checked periodically throughout the duration of the trip. The driver should immediately inform the monitoring base of any abnormalities.

Parking. For parking for meals and the like, the specific public parking areas for heavy goods should be preferred; parking should always be in the open air. The driver should take care and avoid obvious hazards, such as overhead power lines or liquid oxygen tankers when choosing a parking place.

Breakdown. The driver should use all available warning means, such as flashers, reflective triangles, and flashing amber lights, in the case of a breakdown on a highway. No hot work by any nonauthorized persons can be allowed on a liquid hydrogen tanker, unless purged and inerted, and a permit to work has been issued.

Product transfer into customer storage facilities. Transfer operation is only allowed by authorized, trained, and certified personnel of the customer. No transfill operation is allowed during a thunderstorm. The driver should wear all personal protective equipment (gloves, eye protection, helmet, overalls, and protective footwear) during the transfer operation. After product transfer, the delivery hose should be purged of hydrogen before disconnection.

13.4.2.2 Tank Container Transport by Railway

Tank containers transported by railway should be in accordance with RID, *Règlement Concernant le Transport International Ferroviaire des Marchandises Dangereuses* (Regulation Concerning the Transport of Dangerous Goods by International Railway). Only approved tank containers should be used for that purpose and the nitrogen shield vessel should be full. The hydrogen containers should be positioned away from incompatible substances, such as oxidizing agents, and even away from other trains parked nearby in marshalling yards.

The national railway authority should have agreed on the planned route and where the route crosses borders relevant information should be passed to the next railway authority. Detailed written instructions should be given to the national railway authority and all other concerned persons (e.g., emergency services) along the journey route. The hydrogen containers should be fully checked at the end of the journey and the results should be recorded on a checklist.

13.4.2.3 Transport by Waterways and Sea

The transportation of liquid hydrogen by waterways and sea should be in accordance with the International Maritime Organization's International Maritime Dangerous Goods Code and the guidelines given in the EIGA's IGC Doc 41/89. The latter should also be applied to road tankers.

In addition to the requirements of the International Maritime Organization's emergency procedures, specific instructions should be provided to the pertinent shipping authorities. A checklist should also be supplied to ensure that performance monitoring of the hydrogen tank is recorded at regular intervals.

References

1. Biennial Report on Hydrogen Safety, HySafe (Safety of Hydrogen as an Energy Carrier), Chap. VI. http://www.hysafe.org.
2. American National Standard Institute (ANSI), *Guide to Safety of Hydrogen and Hydrogen Systems,* American Institute of Aeronautics and Astronautics, ANSI/AIAA G-095-2004, Chap. 4. ANSI, Washington, D.C., 2004.
3. U.S. Department of Transportation (US DOT), *Guidelines for Use of Hydrogen Fuel in Commercial Vehicles,* Final Report, US DOT, Washington, D.C., 2007.
4. *Code of Practice on Safety in Storage, Handling and Distribution of Liquid Hydrogen,* European Industrial Gases Association, 2002.
5. Rigas, F., and Sklavounos, S., Hydrogen safety, in *Hydrogen Fuel: Production, Transport and Storage,* CRC Press, Taylor & Francis, Boca Raton, FL, 2008, 563–565.
6. Sklavounos, S., and Rigas, F., Computer simulation of shock waves transmission in obstructed terrains, *Journal of Loss Prevention in the Process Industries,* 17, 407, 2004.
7. Health and Safety Executive (HSE), *Explosion Hazard Assessment: A Study of the Feasibility and Benefits of Extending Current HSE Methodology to Take Account of Blast Sheltering,* Teport HSL/2001/04, Sheffield, UK, 2001.
8. Rigas, F., and Sklavounos, S., Evaluation of hazards associated with hydrogen storage facilities, *International Journal of Hydrogen Energy,* 30, 1501, 2005.
9. International Atomic Energy Agency, *Hydrogen as an Energy Carrier and Its Production by Nuclear Power,* IAEA-TECDOC-1085, 1999.
10. National Fire Protection Association, *Vehicular Fuel Systems Code,* National Fire Protection Association, Quincy, Massachusetts, 2006.

14

Conclusion

The prosperity of the present-day developed world has been based on the abundance of fossil fuels as energy source and carrier. Yet these reserves of stored solar energy are finite and, in addition, are mainly composed of carbon. Thus, even if fossil fuel deposits were unlimited, they should be abandoned in the next decades, owing to the massive release of carbon dioxide in the atmosphere and the resulting greenhouse effect, subsequently leading to climate change. Herein lies the necessity for new energy sources and carriers that are based on renewable energy sources such as solar energy. In the opinion of many authors, if the ultimate goal is to develop a sustainable energy economy, the most promising candidate to play the role of storage and transport medium in the near future appears to be hydrogen, provided it will be produced from renewable energy sources. In addition, hydrogen is mainly stored on earth in water of any form, whereas its combustion produces again water; therefore, the cycle closes without environmental deficits of any kind.

Nevertheless, the anticipated entrance to the hydrogen economy has raised many concerns as regards its safe production, transport, storage, and use, among them environmental concerns. Although it is true that hydrogen has been used safely for many decades, this has occurred until now in activities mainly in the chemical industry, where skillful and highly trained personnel are engaged. No one is certain what would happen when a layman handles a potentially hazardous material such as liquid hydrogen to refuel our cars. Such thoughts are sustained by the fact that there is still a significant shortage of knowledge on hazardous properties of hydrogen, for instance on ignition chemistry, deflagration-to-detonation transition (DDT), storage materials, compatibility with other materials, and under extreme conditions of storage and use. Even accident prevention and mitigation measures are not definitely determined yet, and suggested active measures, such as forced ventilation or water mists, may not be as efficient as expected, even if they do not aggravate the situation. Skepticism is also emerging on the environmental effects of large hydrogen leaks if hydrogen comes into extensive use worldwide.

This book is a contribution to basic knowledge on hazardous properties of hydrogen in view of the anticipated hydrogen economy and aims at offering a comprehensive overview on the safe handling and storage of this new energy carrier.

Index

D

Q

T